Noise and the Design of Buildings and Services

This book is based on the conference of the same title, organised in July 1980 by the Construction Industry Conference Centre Limited in conjunction with the Royal Institute of British Architects, the Chartered Institution of Building Services, the Institute of Acoustics and the Institution of Public Health Engineers.

TECHNICAL ORGANISING COMMITTEE

Chairman: Derek J. Croome, BSc, MSc, PhD, CEng, FIOA, MInstE, MInstP, FCIBS, MASHRAE (representing the Chartered Institution of Building Services)

Paul Darrington, ARIBA, FCIArb (representing the Royal Institute of British Architects)

Brian Day, MSc, MInstP, FIOA (representing the Institute of Acoustics)

R.E. Dickson, MIPHE, MRSH, FIOP (representing the Institution of Public Health Engineers)

A.F.C. Sherratt, BSc, PhD, CEng, FIMechE, FCIBS, FInstR

Noise and the Design of Buildings and Services

Edited by

Derek J. Croome
BSc, MSc, PhD, CEng, FIOA, MInstE, MinstP, FCIBS, MASHRAE
University of Bath

Construction Press
LONDON AND NEW YORK

The opinions expressed in this book are the authors' and not necessarily those of the editor, publisher, or the Construction Industry Conference Centre Limited.

Construction Press
Longman House
Burnt Mill, Harlow, Essex, UK

A division of Longman Group Limited, London

Published in the United States of America by
Longman Inc., New York

First published 1982

British Library Cataloguing in Publication Data

Noise and the design of buildings and services.
1. Noise control – Congresses
2. Environmental engineering – Congresses
I. Croome, Derek J.
693.8'34 TH1725

ISBN 0-86095-887-9

Library of Congress Cataloguing in Publication Data

Main entry under title.

Noise and the design of buildings and services.

"Based on the conference of the same title, organized in July 1980 by the Construction Industry Conference Centre Limited in conjunction with the Royal Institute of British Architects . . . [et al.]" ––Prelim. p.
Bibliography: p
Includes index.
1. Soundproofing. 2. Noise control. I. Croome, Derek J. II. Construction Industry Conference Centre. III. Royal Institute of British Architects.
TH1725.N64 693.8'34 81-12631
ISBN 0-86095-877-9 AACR2

Typeset by Lonsdale Typesetting Services, Lancaster and printed in Great Britain by Pitman Press Ltd, Bath

Contents

CONTRIBUTORS

Raymond Andrews, MBE, MArch, DipLArch, DipTP, RIBA, Vice-President Royal Institute of British Architects; Partner, Andrews, Downie & Kelly, London.

Michael M. Barron, MA, PhD, MIOA, Research Associate, Department of Architecture, University of Cambridge.

Michael Bryan, BEng, PhD, Reader, Audiology Group, University of Salford.

Professor Brian L. Clarkson, BSc, PhD, FRAeS, FSEE, FIOA, President, Institute of Acoustics; Dean of Engineering, Institute of Sound and Vibration Research, University of Southampton.

Derek J. Croome, BSc, MSc, PhD, CEng, FIOA, MInstE, MInstP, FCIBS, MASHRAE, Senior Lecturer, School of Architecture and Building Engineering, University of Bath; Associate Partner, Buro Happold (UK).

Brian Day, MSc, MInstP, FIOA, Lecturer in Architecture, University of Bristol.

Stephen Edwards, MSc, CEng, FCIBS, MInstE, AMInstR, MConsE, Partner, Hoare, Lea & Partners, Bristol.

Ian H. Flindell, BSc, MSc, Institute of Sound and Vibration Research, University of Southampton.

Professor John B. Large, BSc, MS, FIOA, MBAC, Director and Chairman of Applied Acoustics, Institute of Sound and Vibration Research, University of Southampton.

David O. Lloyd, AKC, BSc, FIPHE, FICE, FIWES, President, Institution of Public Health Engineers; Partner, Balfours Consulting Engineers, London.

Robin K. Mackenzie, BSc, MSc, PhD, MCIBS, FIOA, Senior Lecturer in Building, Department of Building, Heriot-Watt University, Edinburgh.

Andrew H. Middleton, MSc, FIOA, Technical Manager, Wolfson Unit for Noise and Vibration Control, Institute of Sound and Vibration Research, University of Southampton.

Tim L. Redmore, BEng, MSc, Contract Research Officer, Department of Construction and Environmental Health, University of Aston in Birmingham.

David A. Russell, MA (Oxon), FCIBS, President, Chartered Institution of Building Services; Chairman, Rosser & Russell Ltd., London.

Ron T. Simpson, BArch, RIBA, Architect, Yorke Rosenberg Mardall, London.

Neil Spring, BSc, MInstP, MIEE, FIOA, Partner, Sandy Brown Associates, London.

Rupert Taylor, MIOA, Managing Director, Rupert Taylor & Partners Ltd., London.

John G. Walker, BSc, PhD, MInstP, FIOA, Institute of Sound and Vibration Research, University of Southampton.

Note: the above list was correct in 1980.

Preface

Each person has a noise aura whether at home, walking, travelling or in buildings at work or for pleasure. The way we walk largely determines the sounds from footsteps; on a finer level even the clothes we wear create sounds, just as breezes do blowing through trees. The way we drive influences the noise emission for a particular car. Social habits influence things, too. For example, it is not uncommon to see young people carrying radios around tuned to their favourite pop programme, even when walking in the beautiful countryside. It is not surprising that building design has become increasingly influenced by noise considerations.

Increasing population, travel and technology mean there are many more unwanted sounds from which to protect people today. Mass, damping and stiffness, apart from control of noise at source, are major factors to be taken into account when selecting materials and laying out spaces, to avoid intrusion of unwanted sounds. Selection of equipment, planning and management of the building when in use are other key factors in noise control.

This book forms the proceedings for a conference, 'Noise and the design of buildings and services', held at the Imperial College of Science and Technology in London in July 1980. It is hoped that by involving all the delegates, whether by formal papers or by contributions to discussion, a balanced view has been presented of the problems and some of the solutions facing building designers, whether architects or engineers, covering all design aspects of building acoustics.

Derek J. Croome
Bath, 1981

ACKNOWLEDGEMENTS

The Editor and Technical Organising Comittee wish to thank all the people who have participated in the arrangement of the conference and in the production of this book.

A particular word of thanks is due to Mrs Jean Stephens, Mrs Ruth Brett and their colleagues at the Construction Industry Conference Centre Limited for their efficiency in the organisation and administration.

1 The effects of noise on man

Michael Bryan

SYNOPSIS

The effects of noise on annoyance, communication, hearing, task performance and health are briefly reviewed.

The best documented areas are those concerned with annoyance, communication and hearing. There is general agreement in these areas, not only to the level of noise required to produce these effects but also to the extent of the reaction caused. Perhaps of greatest concern to an employer is the damage that high noise levels can produce on his workers' hearing. Vigorous activity by unions in bringing personal injury claims for occupational hearing loss has resulted in considerable interest in hearing conservation in industry.

The effects of noise on task performance and upon health are much less well known and frequently disputed; however, there is a growing body of evidence to suggest that not only does noise affect man's ability to function efficiently, but it can also impair his health in the long term.

1. INTRODUCTION

The effects of noise on man can be divided into two categories, (a) psychological and (b) physiological, and, for most purposes, can be considered under the following headings:

1. causing annoyance or dissatisfaction
2. affecting communication
3. causing damage to hearing
4. affecting performance of tasks
5. causing permanent changes in the normal functioning of the human organism, resulting in deterioration in mental and/or physical health.

The breadth of knowledge is not equal in all of these areas. A great deal is known about how noise causes annoyance, affects communication and damages hearing, and there is fair agreement in these areas as to the levels of noise required to produce these effects and the extent of the reaction caused. Much less is known about the effects of noise on task performance and health, and there is still little agreement as to its importance in these areas.

An attempt will be made in this chapter to give a brief outline of the current state of knowledge in each of the above areas and, to some extent, the areas of disagreement when they occur.

2. NOISE ANNOYANCE

The most widespread effect of noise is annoyance or, as the Wilson Noise Committee [1] defined it, 'the resentment we feel at an instrusion into the physical privacy which we have for the moment marked out as our own or into our own thoughts or emotions'. The noise of road traffic is by far the most frequent reported cause of annoyance or dissatisfaction, although aircraft, children, people, trains and factories contribute as well but affect fewer people (see Figure 1.1, based upon the study by the Building Research Station of the effects of road traffic noise carried out in 1976 [2]).

Figure 1.1 Noises heard and causing annoyance from outside dwelling [2]

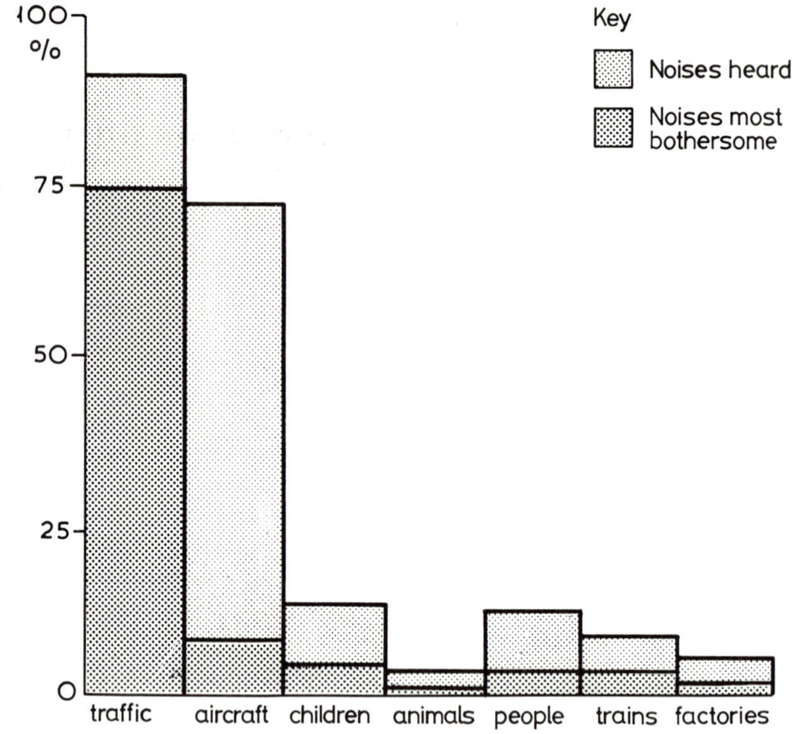

		traffic	aircraft	children	animals	people	trains	factories
Noises heard	(%)	91	72	13	3	12	8	5
Noises most bothersome	(%)	74	8	4	1	3	3	1

It was estimated in 1970 that between '19 and 46 per cent of the UK urban population of 45 million live in roads with traffic flows which produce noise levels likely to be judged undesirable for residential areas'. This statement was taken from a report produced by the Department of Environment Road Research Laboratory [3] which went on to estimate that by 1980 the percentages could rise to between 30 and 61 per cent, owing to the growth of the number of road vehicles alone.

Noise annoyance and the physical and psychological factors which determine its extent have been extensively studied since the early 1930s. From about that time there was not only a rapid increase in the sources of noise, such as vehicular traffic, but also electronic equipment became available which allowed the objective measurement of the levels of noise [4]. Annoyance has been found to depend on the following physical characteristics of the sound:

Intensity [5] – annoyance increases with level
Frequency [5] – high-frequency sounds are more annoying than low-frequency sounds
Duration [6] – annoyance increases with duration
Tonal content [7] – the presence of tones increases annoyance
Localisation [8] – changes of location cause greater annoyance
Distance [9] – distant sources cause greater annoyance than nearer ones
Information content [10] – sounds containing meaningful information, music, voices are more acceptable than pure noises.

The characteristics of the medium of the environment also play a considerable part in determining the degree of annoyance. Some of these are:

Neighbourhood [7] – less noise is tolerated in quieter rural areas than in busy urban areas.
Time of day [6] – noise at night and weekends is more disturbing than daytime weekday noise.
Season of the year [6] – annoyance is greater in the summer when people open windows.

Finally, the characteristic of the individual recipient is the most important factor in determining the extent of annoyance [2]. It has been shown that noise ratings typically correlate with community or average response at about the 0.8 level and can be used for such purposes as planning. However, correlation with dissatisfaction and noise measurement for the individual, as distinct from the average response, is much lower and is typically around the 0.3 level [11]. In other words, it is not possible to predict the annoyance of an individual by noise from a measurement based upon physical parameters of that noise alone. Schultz has pointed out that no physical measurement of noise has yet been devised which will yield correlations with individual response higher than about 0.45 [12]. Moreira and Bryan [13] showed that there were significant stable inter-individual differences in the rating of tape-recorded aircraft, industrial and traffic noise and, while annoyance sensitivity did not appear to be related to age, sex, education and job responsibility, it did correlate significantly with various personality traits. Their work suggested that a noise-sensitive individual 'might typically show a fair amount of empathy and be intelligent and creative'. These findings, though somewhat controversial, have been supported by subsequent studies [14], [15]. However, there is now general agreement that dissatisfaction with environmental noise is strongly dependent upon whether the subject is sensitive or non-sensitive. The effect is shown in Figure 1.2, which shows that dissatisfaction with road traffic noise was dependent upon whether individuals rated themselves as sensitive, neutral or non-sensitive to noise [2].

Figure 1.2 Dissatisfaction with road traffic noise in London for sensitive, neutral and non-sensitive residents [2]
○ sensitive
□ neutral
● non-sensitive

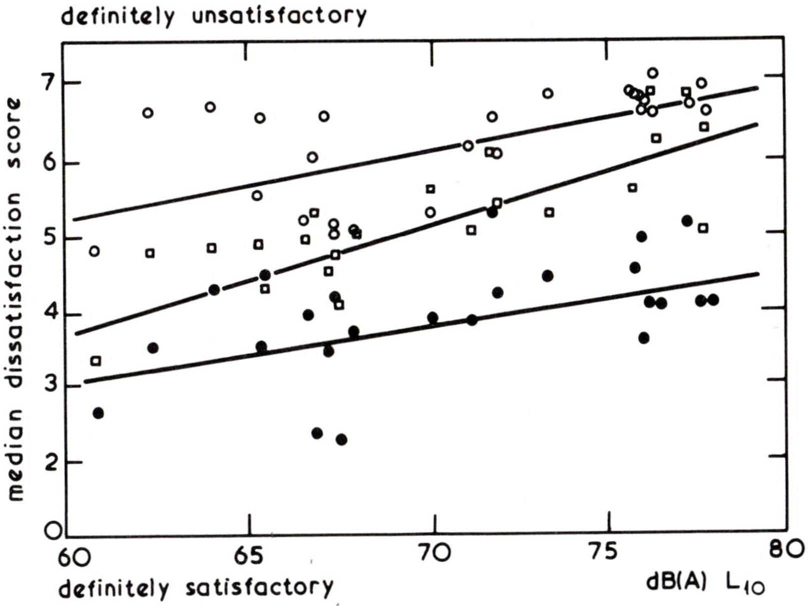

Indeed, much of society's apparently anomalous behaviour towards noise, on the one hand allowing noise levels to rise without check (from transport and entertainment for instance) and on the other protesting vehemently about them, is explicable only if there are noise-sensitive and insensitive members of the community. To obtain a better understanding of annoyance and protection from noise, non-physical parameters associated with the individual must be determined. When it is, eventually, possible to predict most of the variants associated with individual annoyance due to noise, it will be a social-political decision as to how best to use this information to protect the population from noise.

3. THE EFFECTS UPON COMMUNICATION

The ability to be able to communicate easily and without strain is clearly an important factor in the design of the working environment, since most work involves verbal communication to a greater or less extent. The ease with which speech is understood is a major influence upon what noise levels are acceptable in any indoor situation. Also, the type of activity, e.g., use of telephone, lecturing, passing instructions in a production situation, largely determines the level of noise desirable for good intelligibility of speech.

Hearing for speech communication is centred chiefly in the 500, 1000 and 2000 Hz octave frequency bands and any noise in these octave bands will interfere with communication. Therefore, noise measures for speech interference (or Speech Interference Levels, SIL) are based on the average sound-pressure level of the noise in these three octave bands. Maximum speech interference levels for satisfactory lecturing, telephone use and industry are 30, 45 and 70 dB, respectively. Table 1.1 (from Duerden [16]) gives maximum acceptable Speech Interference Levels for a variety of living and working environments.

Table 1.1

Type of room	*Maximum acceptable SIL (dB)*
Small private office	45
Conference room	30–35
Concert hall	25
Typing pool	60
Bedroom	30
Living-room	45
Classroom	30

While speech interference criteria have been traditionally based on the octave frequency bands 500, 1000 and 2000 kHz [17], it has long been known that the consonants contain energy at frequencies as high as 6000 to 8000 Hz. It is becoming increasingly accepted that the frequency range above 2000 Hz is important for good communication, particularly in noisy situations [18]. In 1974 an ISO technical report [19] argued for the redefinition of (Preferred) Speech Interference Level (PSIL), to include the octave band level of 4000 Hz.

4. THE EFFECT OF NOISE UPON HEARING

It was recognised long before preventative measures were available that employment in high noise levels can cause severe and permanent damage to hearing. However, it was only after the Second World War that a concerted effort was made to introduce hearing conservation programmes into those industries with hazardous noise levels. By this time instrumentation was available to measure the noise, and methods for measuring hearing (audiometry) were well established. Surveys of hearing in industry had established the existence of noise deafness, and had suggested maximum permissible exposure levels. Techniques for control of the noise at source were well established and various types of personal hearing protection, such as ear plugs or ear muffs, had been developed during or before the war [4].

Despite this, British industry, as a whole (with the notable exception of some firms active in the 1950s, see for instance *The Annual Report of the Chief Inspector of Factories* for 1954 [20]), has been slow to take up conservation of the hearing of its work force. The publication by the Department of Employment in 1972 of a *Code of Practice for reducing the exposure of employed persons to noise* [21], and a substantial number of successful claims, beginning with that of Berry in 1972 [22], under Common Law for occupational hearing loss, have made it acutely aware of its obligations and very noise-conscious [23].

4.1 Safe exposure levels

The Department's Code of Practice [21] lays down a maximum (equivalent continuous sound level, L_{eq}) for an 8-hour working day of 90 dB(A) and suggests a halving of exposure time for each 3 decibel increase in L_{eq} noise level above this, with a maximum safe limit for any occasion of 135 decibels. Various surveys have suggested this level of 90 dB(A) will only protect 80–90 per cent of the work force (see for instance [24]) and levels of 85 dB(A) or lower give more effective protection. There is a trend in other countries, therefore, to reduce the maximum exposure to 85 dB(A) or lower [25].

4.2 Hazardous industries

It is not known exactly how many industrial workers in the UK are at risk from high-level noise, because no large-scale survey has ever been carried out. It has been estimated, however, that there could be a million or more workers exposed to levels above 90 dB(A) in the manufacturing, *construction*, mining, quarrying, transport and agricultural industries. The incidence among the different occupations is very widespread. Traditionally, the jobs usually associated with noise deafness include blacksmiths, millers, coppersmiths, weavers and spinners, and boilermakers. But the actual spread of occupations involved is extremely broad, as Table 1.2 shows. All the noise measurements in the table have been taken by the author. There may well be other jobs in the various industries mentioned where the workers are exposed to noise levels in excess of 90 dB(A) and are, therefore, at risk.

Table 1.2 Typical hazardous jobs in the common noisy industries (in all cases hearing protection should be worn)

Industry	*Job*	*Typical daily noise dose dB(A) (L_{eq})*
Steel fabrication	Hand hammering	112
Shipbuilding	Chipping	115
	Riveting*	115
	Propeller making	120
	Caulking	118
Aircraft construction	Riveting	111
	Engine testing	120
Papermaking	Production line	97
Textiles	Spinning	94
	Weaving	100
Iron and steel foundries	Fettling	110
Drop forging	Stamping	104
	Blacksmiths	98
Woodworking	Planing machine	93
	Circular saw	93
Locomotive repairing	Riveting*	115
	Chipping	103
Motor vehicle manufacture	Fettling	110
	Engine testing	101
	Press operating	97
Food and drink manufacture	Production line	95
Building construction	Unsilenced road drill	106
	Bulldozer cab	100

* Fortunately now obsolete.

4.3 Noise deafness sensitivity

When a worker is exposed to levels of noise in excess of 85 dB(A), he may become aware after as little as one to two years that his hearing is becoming dull and that he has noises in his ears which, initially, only occur temporarily and disappear on resting. In extreme cases, these noises may become loud and cause considerable distress, and he may have to give up working in noise [26]. On the other hand, he is more likely to be unaware that the damage is being caused to his hearing because this usually occurs in the 4000–8000 Hz region at the upper end of the speech-frequency range. As exposure to noise continues, so the damage to hearing increases and spreads down to the main speech frequencies around 1000–2000 Hz (see Figure 1.3) [27, 28, 29]. The worker initially finds difficulty in understanding speech, particularly under noisy conditions, such as conversations in clubs and at meetings. Not only is his ability to discriminate speech sounds affected but also his directional ability is impaired as well [18]. As the damage progresses, he will have difficulty with normal speech under quiet conditions and, typically, after 35 to 39 years will require a hearing aid, although this has been shown to be of limited benefit [30].

Figure 1.3 Growth of hearing loss due to weaving noise (100 dB(A) (L_{eq})) and age

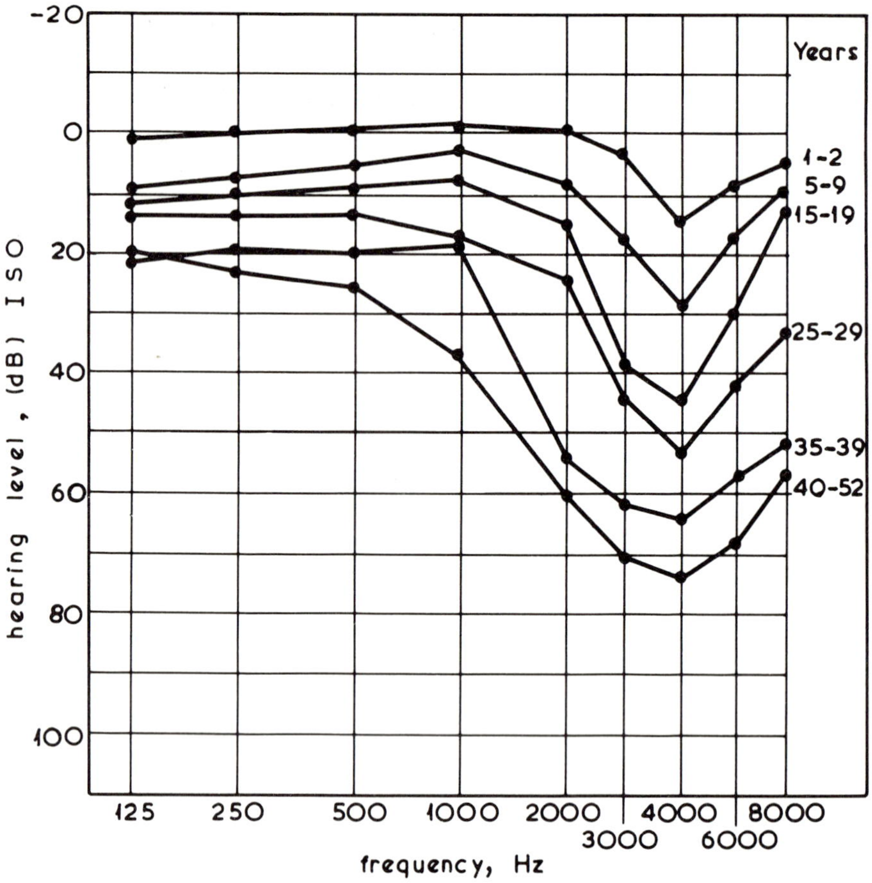

Exposure to noise of these levels causes progressive and permanent destruction to the hair cells in the inner ear, which convert sound waves to electrical impulses that are then carried to the brain. The loss of hearing is usually binaural though in some cases (e.g. weapon firing) there may be some asymmetry in the damage.

There is a wide range in susceptibility to hearing damage, just as there is a wide range in sensitivity to annoyance by noise. The two effects are not known to be related: a person who cannot tolerate noise at any price is not necessarily one who will show considerable hearing damage to modest exposure to high-level noise. As far as is known, sensitivity to annoyance by noise is determined by personality and is psychological in origin, whereas, it would not be unreasonable to suppose that susceptibility to hearing damage is physiological in origin. But, while it is possible to predict noise-annoyance sensitivity fairly well, at present there is no way of predicting sensitivity to hearing damage. If the latter were possible, it would mean that substantial savings to the cost of running a hearing conservation programme could be achieved. Attention could be directed towards protecting the hearing of those workers most likely to suffer hearing damage.

The extent of the range of sensitivity to noise-induced hearing loss is seen in Figure 1.4. The data are based on the work of Burns and Robinson, from which it is possible to calculate the loss expected for a subject of high and one of low susceptibility (2 per cent and 98 per cent levels). This is done in Figure 1.4 for 40 years' exposure at an L_{eq} noise level of 95 dB(A), which is typical of levels found in noisy industry. The figure shows that the subject of low susceptibility has virtually no loss due to noise and, even assuming he has average hearing for his age, he still will have no handicap for understanding conversational speech. The subject of high noise susceptibility has around a 40 dB loss of hearing due to noise alone and, when his hearing level due to age is taken into account as well, he has a substantial hearing loss indeed. It might be concluded that there is little point in providing hearing protection for workers of low susceptibility in a situation like this, and it is of prime importance to protect the hearing of those of high susceptibility, or even not to employ them at all in levels in excess of 85 dB(A).

Figure 1.4 **Hearing loss of two workers exposed to noise of 95 dB(A) for 40 years. One has high susceptibility to noise damage, the other, low.**
(a) noise alone } low susceptibility
(b) noise + age
(c) noise alone } high susceptibility
(d) noise + age

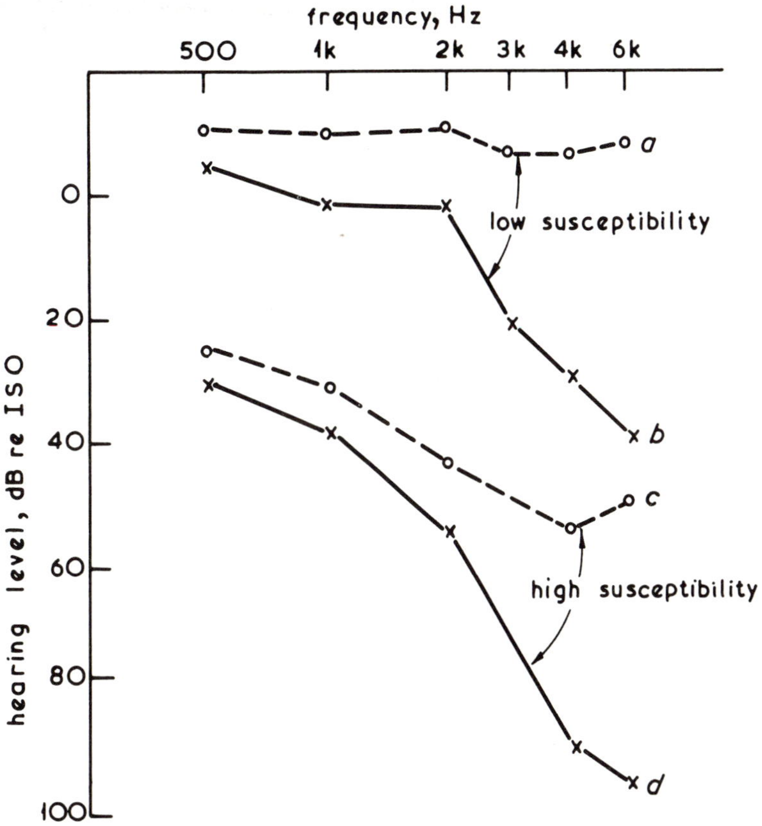

Until quite recently, it was considered that the amount of permanent hearing loss (p.t.s.) a worker would sustain when exposed to noise over long periods was related to, and could be predicted from, the temporary threshold shift (t.t.s.) that he would sustain when exposed to that same noise for a day. The NPL/MRC survey gave an opportunity to test this hypothesis [31]. However, as Burns [32] points out, the best correlation found between permanent and temporary threshold shift was only 0.38, and this was, unfortunately, too low to be generally useful as a prognostic test for persons proposing to enter a noisy occupation.

The only method currently available for the early detection of those workers most sensitive to hearing damage is by conducting annual hearing tests. It is encouraging that such a practice has become widespread in the UK since the early 1970s [26]. Such a hearing test can detect slight hearing damage before it becomes extensive and noticeable by the worker, and produces a handicap. Appropriate steps can then be taken by the industrial medical officer to ensure that further damage is prevented.

Further medical/physiological research is required to be able to predict noise susceptibility and to differentiate the effects of noise deafness from that of deterioration of hearing due to age (presbycusis).

5. EFFECTS OF NOISE ON TASK PERFORMANCE

Several studies have been carried out in industry which show that for certain tasks (involving vigilance, such as machine-minding processes, e.g., weaving, spinning and papermaking, etc) noise does, apparently, have an adverse effect upon performance [52].

The earliest study of importance concerning the effects of noise on task performance is that of Weston and Adams [33] who showed that, for weavers, a reduction of noise level from 96 to 83 dB produced an overall improvement in output of 1 per cent, which was equivalent to a 12 per cent improvement in workers' efficiency. Broadbent and Little's study [34] found that there was an 80 per cent reduction in the number of rolls of broken film, and shutdowns were 40 per cent fewer, in the cine-film perforation plant, after noise levels were reduced by as little as 10 dB. Kovrigin and Mikheyev [35] found that increasing the noise level, by means of loudspeakers, in a room used for sorting postal letters, caused a systematic increase in the number of errors. The noise levels were varied from 78–80 dB to 95 dB and productivity fell by about 30 per cent from the lowest to the highest level. The workers reported such symptoms as headaches, deterioration of appetite, fatigue, etc. The Russian workers concluded that workers' productivity depended to a significant degree on operative noise levels and, also, that it is advisable to set standards for maximum permissible levels for communication enterprises. Unfortunately, we do not know if these effects would be sustained if the higher noise levels were present continuously.

The results of these field studies have been confirmed in the laboratory in experiments where subjects had to perform vigilance tasks in noise and generally showed a deterioration of performance under such conditions. In other tasks, such as manual work of a repetitive nature, noise seems to have little effect upon performance [36].

The situation is somewhat confusing as to whether or not noise affects high-level intellectual tasks. Some workers have produced data which support this idea, while other workers suggest that moderate noise levels do not affect mental tasks. Some light has been thrown on this divergence of opinion by work done at Salford University [37]. This work showed that noise of 80–90 dB(A) peak had a differential effect upon the performance of students doing IQ tests. Those with high IQ did less work and made more mistakes in the noise while those with low IQ did more work and made fewer mistakes in the same noise. Overall, there was no effect, as the two different effects cancelled one another out. This result would seem to have important implications for workers in many offices, where noise levels have increased substantially due to the introduction of automation and computers. It would seem that a number of office workers would suffer a substantial loss of efficiency in such high noise levels. Such a finding would also be important for management in industry working on, or adjacent to, the shop floor and hence in noise levels up to perhaps 90 dB(A).

Related to both noise annoyance and the effect of noise on task performance were the findings of a laboratory study carried out on Preferred Noise Levels for mental task performance by Bryan and Tolcher in 1976 [38]. They found that there was no deterioration in task performance if subjects were allowed to carry out a variety of mental tasks of differing complexity at their preferred listening level. While there was, not surprisingly, a general trend for difficult tasks (calculating and comprehension) to require lower levels than simple (routine vigilance) tasks, there was an extremely wide range of preferred listening levels. Some subjects, in this laboratory study, preferred to work in conditions of near silence, while others worked and performed quite happily in levels in excess of 90 dB(A). This study demonstrates the importance of taking into account individual differences in response to noise, in any attempt to establish criteria for acceptable noise levels in the work environment.

6. EFFECTS OF NOISE UPON HEALTH

In this section we are concerned with the possible effects that noise may have upon health, using the World Health Organization definition: 'Health is a state of complete physical, mental and social well-being and not merely an absence of disease or infirmity.' Annoyance effects must be included in this section where, following Burns [32], annoyance (by noise) may be regarded as the displeasure or resentment caused by sound, either by its physical presence or because of the implications arising out of its presence. The last clause is important because one of the implications arising out of noise annoyance could be that it is a factor in mental or physical ill-health.

It is useful to start this section with a statement as to what are considered to be acceptable levels of noise in the work environment from an *annoyance* viewpoint. There are *very* few data on this, most writers are concerned only with noise levels for hearing protection. However, Kosten and Van Os have suggested levels of 55 NR (60 dB(A)) for typing offices and 65 NR (70 dB(A)) for workshops respectively [39]. Generally, in the former situation this level is equalled or bettered, while few would deny that only rarely do workshops (and we can include most industrial situations here) come anywhere near this 70 dB(A) criterion. Most industrial work environments probably have noise levels in the range 75 to 85 dB(A) and it must, therefore, be considered that generally industrial noise levels are higher than they ought to be. If the 70 dB(A) level proposed by Kosten and Van Os is realistic, there ought to be some evidence that the health of those workers exposed to levels much over this is being affected.

Jansen [40] found a greater percentage of blood circulation, and heart and equilibrium problems in workers in very noisy industries compared with less noisy industries, as is shown in Table 1.3. The differences in peripheral circulation and heart problems in the two classes of industry were statistically significant. The numbers in brackets are the sample sizes.

Table 1.3 Percentage occurrence (German industrial workers)

	Very noisy industries	*Less noisy industries*
Peripheral circulation problems	62 (410)	49 (165)
Heart problems	26 (161)	16 (50)
Equilibrium problems	20 (128)	14 (51)

Shatalov, Saitanov and Glotova [41] compared cardiovascular characteristics of spinning-mill workers (85–95 dB) with those of ball-bearing plant workers (114–120 dB) and found that the higher noise-level situation apparently had a more deleterious effect upon the cardiovascular system than the quieter one, though even in the spinning mill the noise levels are high, which is, perhaps, reflected in the fact that the percentage of those having depressed T waves after work is high, at 21.2 per cent. Andriukin [42] has presented data which show that the incidence of hypertension (arterial blood pressure) tends to be greater in workers exposed to high-frequency shrill lathe noise, and to the very intense broad-band noise found in ball-bearing production shops, than for men working in less intense noise. This is shown in Table 1.4, which gives the incidence of hypertension in male and female workers, by age groups, in noisy shops.

Table 1.4 Percentage incidence of hypertension

	Below 40 years	*Above 40 years*
Tool-making shops	2.6	12.1
Sorting shops	4.1	16.5
Lathe shops	4.3	18.3
Ball-bearing shops	4.6	17.0

Related to health is the question 'Do these workers in high noise-level situations have a greater absenteeism rate than those working in quieter environments?' Broadbent and Little [34] found a small decrease in absenteeism (time as a percentage of possible hours worked) from 5.18 per cent to 4.43 per cent as a result of reducing the noise level (from 99 to 89 dB) in the perforating bay of a cine-film manufacturing plant. However, as absenteeism also fell in an adjoining untreated bay during the same period, the improvement in attendance might have been due to an increase in morale of the workers because 'something was being done for them' rather than to a decrease in noise level. Ribeiro de Almeida [43] has reported a reduction in absenteeism in a group of 125 workers in an electric punch-card machine office from 1.826 per cent to 1.396 per cent (a reduction of 430 days of absences) as a result of reducing the noise level in the office by an average of 23 dB. He states that this reduction was almost exclusively due to a decrease in neuropsychiatric cases and, presumably, was owing to the decreased noise levels. Ferguson [44] found, in the study of Australian postal workers, that the proportion of neurotic absentees was greater in telegraphists than in clerks, and in clerks than in mechanics. The telegraphists worked in more intense noise (83 dB(A) with peaks of 103 dB(A)) and at high levels of concentration and verbal communication. The only occupational characteristics that could be identified strongly with neurosis were the negative attitudes to job, supervision and noise in the work place.

The implication of the above results is that intense noise is probably responsible for the differences in health reported between workers in relatively quiet and noisy industries. However, the author would concur with Kryter [45] and Croome [52] 'that the presence of other possible important factors in the (noisy) work situations must not be overlooked: (a) poor ventilation, heat and light; (b) danger from accidents; (c) anxiety over job security; and (d) personnel selection (on the average those men in the noisier jobs were perhaps less well off in terms of general health and economic and social status, prior to and during employment, than were men in less noisy occupations)'.

Since about 1970 a good deal of evidence has been reported suggesting that environmental noise can be deleterious to both mental and physical health. The earliest work was that of Abey-Wickrama *et al.* [46] who found a greater incidence of mental hospital admissions close to London Airport than in areas further away, where the noise levels from the aircraft were lower. Gattoni and Tarnopolsky [47] repeated the study in 1973 and, while they could not confirm the highly significant differences shown in the earlier work, did find a trend in agreement with the original findings. Reference should also be made to the Polish-American study [53] discussed by MacKenzie in Chapter 8.

The whole area of the relationship between environmental noise and health is one of some controversy. On the one hand, a conservative view is taken by some authors. See, for instance, the excellent review by McLean and Tarnopolsky [48] published in 1977. On the other hand, fairly extreme claims are made concerning the effects of noise on health. Nowell Jones and Tauscher [49] found a higher incidence of reportable birth defects for the years 1970, 1971 and 1972 within the 90 dB(A) contours around Los Angeles airport than in the rest of the country. Meecham and Shaw [50] reported a substantial increase in mortality rates around the same airport in 1979. Also Ettema *et al.* [51], in a very extensive study around Schiphol Airport in the Netherlands, found strong evidence of the adverse effects of aircraft noise on health. There was an increase in hypertension and in cardiovascular impairment, and of visits to the general practitioner, not only for psychosomatic symptoms, but also for cardiovascular diseases. They also found greater use of sedative and hypnotic drugs. Their study also showed that traffic noise and textile noise were likely to cause an increase in hypertension and strongly suggested that long-term exposure to noise is a risk factor for cardiovascular disease in both daily living and working conditions.

REFERENCES

1. *Noise, Final Report*, Cmnd 2056, HMSO (1963).
2. Langdon, F.J. and Buller, I.B. *The effects of road traffic noise in residential areas*, Building Research Establishment, Current Paper CP10/77 (1977).
3. Department of Environment, Report LR 357. *A review of road traffic noise*, Road Research Laboratories, Berks (1970).
4. Tempest, W. and Bryan, M.E. 'Noise measurement and evaluation', *Proc. IEE*, No. 10R, Vol. 123; *IEE Reviews*, 1113–37 (1976).
5. Laird, D.A. and Coye, K. 'Psychological measurement of annoyance as related to pitch and loudness', *J. Acoust. Soc. Am.*, **28**, 872–7 (1956).
6. Stevens, K.N., Rosenblith, W.A. and Bolt, R.H. 'A community reaction to noise: Can it be forecast?', *Noise Control*, **1**, 63–71 (1955).
7. British Standard 4142, *Method of rating industrial noise affecting residential and industrial areas*, British Standards Institution (1967), with AMD 1661 (1975).
8. Harris, C.M. (Ed.). *Handbook of noise control*, McGraw Hill, New York (1967).
9. Bowsher, J.M., Johnson, D.R. and Robinson, D.W. 'A further experiment in judging the noisiness of aircraft in flight', *Acustica*, **17**, 245–67 (1966).
10. Kosten, D.W. and Van Os, G.J. *Community reaction criteria for external noises*, National Physical Laboratory Symposium on the Control of Noise, HMSO, London (1962).
11. Griffiths, I.D. and Langdon, F.J. 'Subjective response to road traffic noise', *J. Sound*

and Vib., **8**, 16–32 (1968).
12. Schultz, T.J. *Community noise ratings*, Applied Science Publishers, London (1972).
13. Moreira, N.M. and Bryan, M.E. 'Noise annoyance susceptibility', *J. Sound and Vib.*, **21**, 449–62 (1972).
14. Bryan, M.E. 'Prediction of noise annoyance – 1 Noise susceptibility; evidence from field studies', *Proc. British Acoust. Soc.*, Vol. 2 (3) (1973).
15. Templeton, D. 'An investigation into the noise climate around two motor race circuits and subjective response of local residents', M.Sc. thesis, University of Salford (1979).
16. Duerden, C. *Noise abatement*, Butterworth, London (1972).
17. Webster, J.C. 'S.I.L. – past, present and future', *Sound and Vibration*, **3**, No. 8 (1969).
18. Bryan, M.E. 'Hearing handicap and noise induced deafness', Paper read at The Third British Conference on Audiology, Manchester, July 1979.
19. ISO Recommendation R3352, *Assessment of noise with respect to its effect on the intelligibility of speech*, International Organisation for Standardisation, Geneva, Switzerland (1974).
20. *The Annual Report of the Chief Inspector of Factories*, Cmnd 9605, HMSO, London (1954).
21. Department of Employment. *Code of Practice for reducing the exposure of employed persons to noise*, HMSO, London (1972).
22. 'Berry vs Stone Manganese Ltd', *Knights Industrial Reports*, **12**, 13–35 (1972).
23. Tempest, W. 'Medico-legal aspects of noise-induced hearing loss', in *Disorders of auditory function* II (Ed. S.D.G. Stephens), Academic Press, London (1976).
24. ISO R1999, *Assessment of occupational noise exposure for hearing conservation purposes*, International Organisation for Standardisation, Geneva, Switzerland (1975).
25. Hassall, J.R. and Zaveri, K. *Acoustical noise measurements*, Bruel and Kjaer, Denmark (1979).
26. Bryan, M.E. and Tempest, W. *Industrial audiometry* (2nd ed), Bryan and Tempest (Noise Consultants), Kismet, Croyde Road, St Annes, Lancs (1980).
27. Bryan, M.E. and Tempest, W. *Examples of industrial audiograms*, Bryan and Tempest (Noise Consultants), Kismet, Croyde Road, St Annes, Lancs (1978).
28. Taylor, W., Pearson, J., Mair, A. and Burns, W. 'Study of noise and hearing in jute weaving', *J. Acoust. Soc. Am.*, **38**, 113–20 (1965).
29. Pearson, J.C.G. 'Prediction of presbycusis', *J. Soc. Occup. Med.*, **27**, 125–33 (1977).
30. Kell, W.L., Pearson, J.C.G., Acton, W.I. and Taylor, W. 'Social effects of hearing loss due to weaving noise', in *Occupational hearing loss* (ed. Robinson, D.W.), Academic Press, London (1971).
31. Burns, W. and Robinson, D.W. *Hearing and noise in industry*, HMSO, London (1970).
32. Burns, W. *Noise and man* (2nd ed), Murray, London (1973).
33. Weston, C. and Adams, S. *The performance of weavers under varying conditions of noise*, Report No. 70, Industrial Health Research Board, HMSO, London (1935).
34. Broadbent, D.E. and Little, A.A.J. 'Effects of noise reduction in a working situation', *Occup. Psychol.*, **34**, 133–40 (1960).
35. Kovrigin, S.D. and Mikheyev, A.P. *The effects of noise level on working efficiency*, Report N 65-28297, Joint Publications Research Service, Washington DC (1965).
36. Broadbent, D.E. Chapter 22 in *Experimental psychology in industry* (ed. Holding, D.H.), Penguin Books, Harmondsworth (1969).
37. Bryan, M.E. 'Does noise handicap the intelligent?', *Physics Bulletin*, **24**, 89–91 (1973).
38. Bryan, M.E. and Tolcher, D. 'Preferred noise level whilst carrying out mental tasks', *J. Sound and Vib.*, **45**, 139–56 (1976).
39. Kosten, C.W. and Van Os, G.J. *Community reaction criteria for external noises*, National Physical Laboratory Symposium No. 12, HMSO, London (1962).
40. Jansen, G. 'Adverse effects of noise on iron and steel workers', *Stahl. Eisen.*, **81**, 217–20 (1961).
41. Shatolova, N.N. Saitanov, A.O. and Glotova, K.V. *On the state of the cardiovascular system under conditions of exposure to continuous noise*, Report T-411-R, N65-15577, Defense Research Board, Toronto, Canada (1962).
42. Andriukin, A.A. 'Influence of sound stimulation on the development of hypertension, clinical and experimental results', *Cor. Vasa.*, **3**, 285–93 (1961).

43. Ribeiro de Almeida, H. 'Influence of electric card punch machines on the human ear', *Arch. Otolaryngol.*, **51**, 215–22 (1950).
44. Ferguson, D. 'A study of occupational stress and health', *Ergonomics*, **16**, 649–63 (1973).
45. Kryter, K. *Effects of noise on man*, Academic Press, New York (1970).
46. Abey-Wickrama, I., a'Brook, M.F., Gattoni, F.E.G. and Herridge, C.F. 'Mental hospital admissions and aircraft noise', *Lancet*, ii, 1275–7 (1969).
47. Gattoni, F. and Tarnopolsky, A. 'Aircraft noise and psychiatric morbidity', *Psychological Medicine*, **3**, 516–20 (1973).
48. McLean, E.K. and Tarnopolsky, A. 'Noise discomfort and mental health', *Psychological Medicine*, **7**, 19–62 (1977).
49. Nowell Jones, F. and Tauscher, J. 'Residence under an airport landing pattern as a factor in teratism', *Archives of Environmental Health*, 10–12 (Jan/Feb, 1978).
50. Meecham, W.C. and Shaw, N. 'Effects of jet noise on mortality rates', *British Journal of Audiology*, **13**, 77–80 (1979).
51. Ettema, J.H. and Zielhuis, R.L. 'Health effects of exposure to noise, particularly aircraft noise', *Int. Arch. Environmental Health*, **40**, 163–207 (1977). (Nine papers on various aspects of noise and health with various authors.)
52. Croome, D.J. *Noise, buildings and people*, Pergamon (1977).
53. *The influence of materials and construction on the aerobatic climate in dwellings and its effects on residents' health*, Ministry of Building, Poland (1975).

2 Environmental noise criteria

John B. Large, Ian H. Flindell and John G. Walker

1. INTRODUCTION

This chapter sets out to describe a selection of environmental noise criteria for use by building designers, and to examine the rationale behind the development of those criteria. It is not concerned with the assessment of the range of noise levels which can cause damage to hearing, but with those levels of environmental noise which can cause activity disturbance or interference with relaxation, sleep or communication. Furthermore and, perhaps, more importantly, it is concerned with perceived environmental quality.

In the United Kingdom considerable effort has been expended on formulating methods for quantifying levels of environmental noise. This has resulted in the adoption of a variety of noise scales and indices for use in evaluating environments over a wide range of different conditions. For example, the Noise and Number Index (NNI) is used for aircraft noise, the L_{10} (18 hour) for road traffic noise, the Corrected Noise Level (CNL) for noise from industrial premises and other fixed installations, and L_{eq} dB(A) for railway noise, construction site noise and noise abatement zones. The Noise Criteria and Noise Rating curves are often used for the assessment and design of indoor environments.

However, this proliferation of different noise indices and scales for use in different situations and with different noise sources can *and does* lead to confusion and difficulties, especially in the valuation of complex environments resulting from a number of noise sources. The concept of the unique dose-response relationship seems attractive but it has not, to date, been employed with particular success. There are also problems caused by the necessity to balance the benefits of ideal noise criterion levels against the costs of noise-reduction measures. Thus, those individuals or communities which may be especially sensitive to noise can only expect degrees of protection which would produce an acceptable environment for the majority.

2. RATIONALE

Environmental noise criteria are specified levels of noise, expressed in terms of scales or indices, which are selected to define degrees of acceptability, or to define appropriate use or design. Any criterion must use a scale or index which can be measured or predicted sensibly, and they are defined in terms of a specified function of sound-pressure level, expressed in decibels (dB), and variation with time. A noise scale may be used to describe a non-varying noise, or the peak level of a noise event, and is obtained by applying some form of frequency weighting to the measuring instrument. Examples are the dB(A), dB(B) and dB(C) scales. A noise index includes an allowance for variations of noise level with time, and may also be dependent on the situation; for example, the level of background noise could be included, or the time of day or night could be allowed for, with a weighting factor.

In order to satisfy the demands of objectivity, noise indices and scales must be reliable and repeatable. Various standards have been laid down as to the required precision of measuring instruments (see, for example, BS 4197: 1967 [1]). The criterion should state precisely (or by direct reference to published standards) the procedure to be adopted for measurements or predictions, and the procedures must be applied by competent staff.

Any noise index or scale used in a practicable criterion must be economically measurable with existing technology. This constraint is gradually being relaxed with the introduction of microprocessors, but the long-term averaging of noise levels to yield indices such as the 24 hour L_{eq} can still require large amounts of manpower, as no system yet devised can reliably identify contributions to the overall environment from any one particular source, without the aid of the noise expert's ear.

The degree of acceptability, or definition of appropriate use or design, must be obtained by measurement, or inferred from experience. Measurements of degrees of acceptability can be obtained from social surveys, case histories, complaint studies, behavioural observations and laboratory studies. Each research method must be applied in conjunction with objective measurements or predictions of noise levels, in order to enable criteria to be defined. It is often necessary to amass large amounts of data in such a programme, as the best method of measurement of both the noise levels and their effects cannot be known until the end of the study. It is recommended that some form of standardised research procedure should be adopted in future [2] in order to be able to make meaningful comparisons between studies. These recommendations include the use of L_{eq} as an objective noise definition for all studies, the use of a core of standardised questionnaire items in social surveys and laboratory studies of annoyance, the development of methods for combining the contributions from laboratory and field studies, and the inclusion of a wide range of acoustical and locational variables in each study.

There are a number of effects of noise on man which could be measured in order to define criteria in terms of degrees of acceptability. As well as being able to cause a reduction in perceived environmental quality, noise can have a physiological effect on the auditory and non-auditory systems, and it can interfere directly with relaxation, sleep, communication, or task performance. In a review of these effects Bastenier, Klosterkoetter and Large [3] were able to draw the following conclusions.

Autonomic responses. Although a great deal of work has been reported concerning cardiovascular, respiratory, pupil dilation, skin resistance and hormonal responses to noise stimuli in laboratory situations, it is extremely difficult to use the data to define degrees of acceptability, as it is not known whether any particular autonomic responses are potentially harmful and, indeed, whether the same responses occur in real life. There is a limited amount of evidence for increased rates of foetal abnormality and cardiovascular disease, for example near airports, but it is very difficult to conclude that these effects are caused solely by noise, or what part noise plays. This type of response is of limited value for the selection of environmental noise criteria.

Sleep disturbance. Noise can delay or even prevent sleep by keeping arousal levels too high, and it can also alter cyclic sleep patterns by causing changes in sleep stage. Generally speaking, sleep disturbance does not occur when the indoor night-time equivalent noise level is below 35 dB(A) and the average maximum level of peak noise events does not exceed the equivalent continuous noise level by more than 10 dB. However, awakening thresholds depend to a great extent on a number of non-acoustic variables and, furthermore, there is no conclusive evidence on the extent to which prolonged disturbance to sleep patterns can affect health.

Noise and relaxation. The effects of noise on rest and relaxation are very difficult to quantify. The individual state of mind plays an important role in determining the degree of disturbance but for rest and relaxation associated with sick and convalescent persons daytime equivalent sound levels of 45 dB(A) would appear to provide an adequate indoor environment. At night the levels ought to be 5 to 10 dB lower.

Task interference. These effects have been investigated mainly in laboratory experiments. There appears to be no significant interference with work or other activities which do not depend on the assimilation of acoustic information, if the steady noise level is below 90 dB(A). Tasks that require a high degree of concentration are more readily disturbed. Impulsive noise can cause significant task interference if it startles, and causes arousal of the central and autonomic nervous system. However, the levels of noise that have been demonstrated to cause task interference are much higher than the levels that would be considered desirable from other points of view. Thus, task interference cannot form a basis on which to select environmental noise criteria.

Speech interference. High levels of speech intelligibility are essential for reliable communication. Speech intelligibility is a function of normal or raised voice level, distance between communicants and background noise, and is quoted as a percentage of correctly understood syllables. (Speech interference, on the other hand, is the percentage of those incorrectly understood.) Figures 2.1 and 2.2 [4] illustrate these functional relationships. Good speech intelligibility can be achieved outdoors at one metre distance, with a background noise level of 65 dB(A) equivalent continuous sound level (L_{eq}) or less. This is assuming normal conversational levels. Reduced conversation levels, such as may be encountered during a medical consultation, or tutorials, require background noise levels of 5 to 10 dB lower, in order to maintain the same levels of speech intelligibility. In situations where the noise includes large variations in level, for example near an airport, motorway or railway, then a lower level of background noise expressed in terms of L_{eq} is also necessary for clear speech recognition. All these points also apply to conditions for viewing television, listening to radio or participating in a telephone conversation.

The importance of good conditions for speech intelligibility is borne out by the fact that speech interference is often mentioned as a particular disturbance in social surveys.

It can be seen that physiological responses to, and activity interference caused by, noise do not form relationships that can easily be adapted to select generalised environmental noise criteria. Task interference results only from what would otherwise be considered excessive levels, and thresholds for minimum or zero sleep disturbance are so low that it is not always possible to achieve the requisite levels within typical urban environments. If people wish to have the benefits of industrialisation, then, with current technology, there is an economic limit to the amount of noise reduction possible, which defines the levels of noise which those people may have to put up with, irrespective of the acceptability of those noise levels in any absolute sense. As regards speech interference, it

Figure 2.1 Percentage sentence interference as a function of outdoor equivalent sound level, L_{eq}, in dB(A) [4]

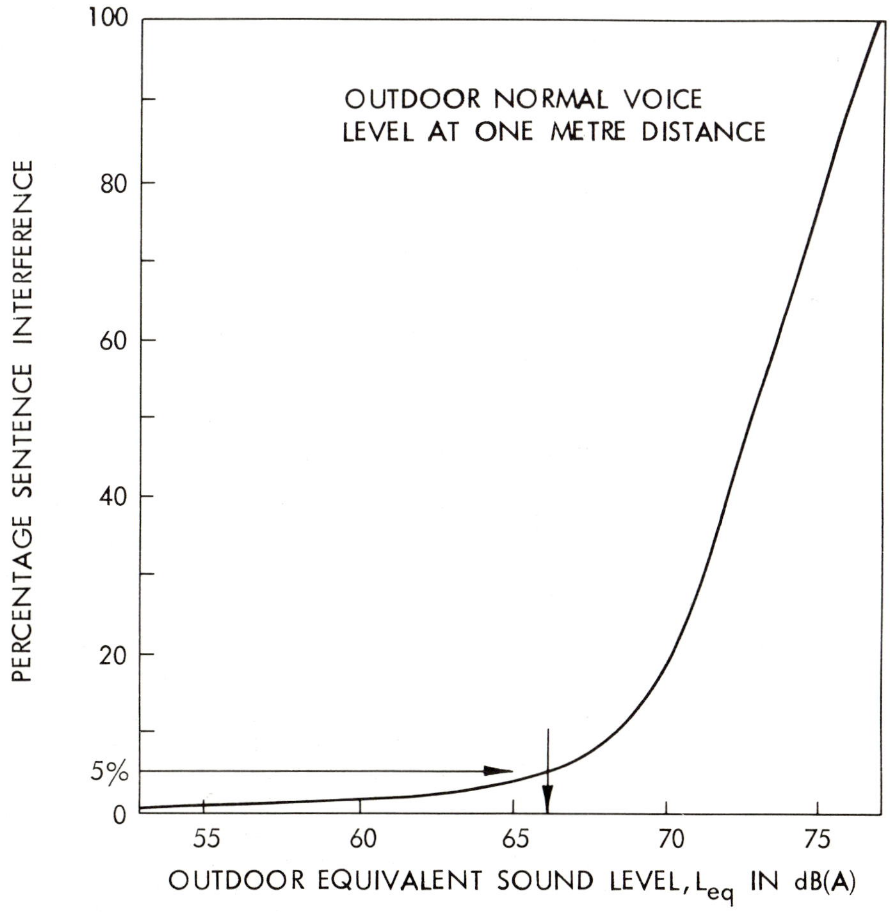

Figure 2.2 Sentence intelligibility as a function of sound level for normal voice in indoor situation. Relaxed conversation would be 10 dB lower [4]

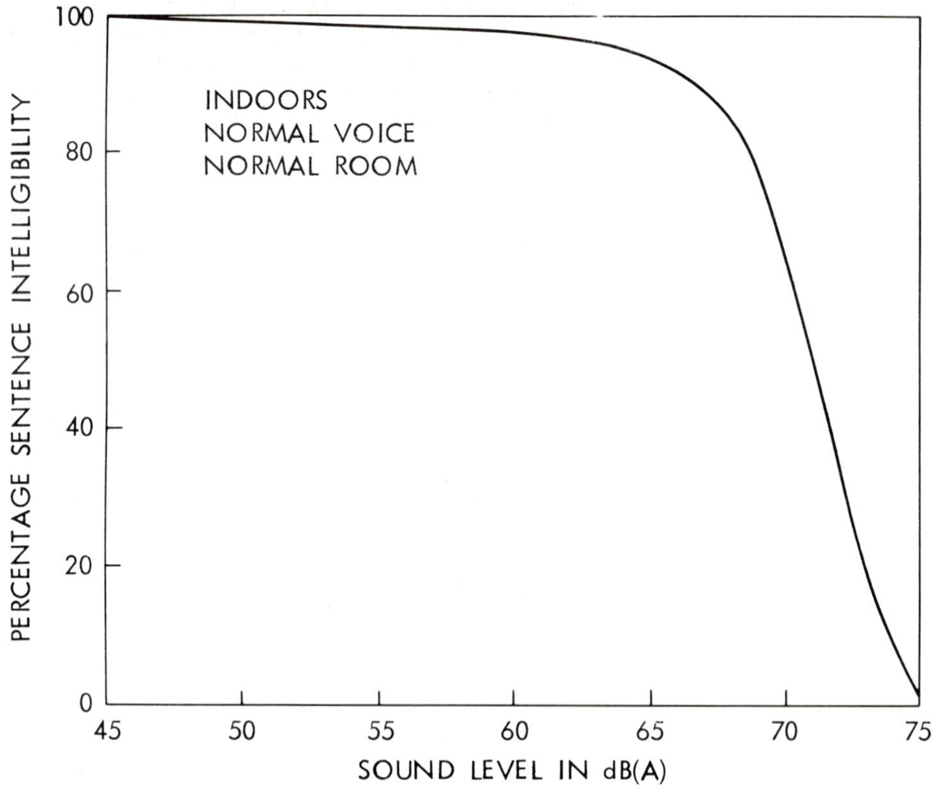

is difficult to select levels on the basis of acceptability, since in different situations people are accustomed to employing different speech levels in order to achieve whatever level of intelligibility they desire. Thus, the above effects could only serve to define noise criteria under specific and narrow circumstances, and are not suitable for the selection of environmental noise criteria.

Perceived environmental quality is an important aspect of the selection of noise criteria. It depends upon human value judgements and is something which can be measured subjectively, but has no absolute determinants. It is, also, a factor which can be evaluated on a continuous scale, varying in relative terms from 'unacceptable' to 'acceptable' or even 'desirable'. It offers an opportunity for trading off the subjective effects of environmental noise against the costs of noise-reduction measures and thus provides a sound basis for the administrative procedure of setting criteria. It provides each individual with the opportunity to make his own assessment or subjective weighting of the relative importance of each type of noise nuisance or disturbance in forming a judgement of perceived environmental quality.

The most convenient way of measuring perceived environmental quality is to examine annoyance responses by means of questionnaires, either in field surveys or laboratory studies, or both. The view that is sometimes held that annoyance responses are little more than fickle responses of cranks or unstable persons is easily dealt with by reference to the large body of data showing consistent increases of degree of reported annoyance with noise level. Degrees of annoyance can be measured by using a scale with labelled end points at 'no annoyance' and 'extreme annoyance', and depend upon the unannoyed or marginally annoyed person considering for himself the sorts of situation where he might react with extreme annoyance. Survey data have shown only mild relationships, if any, between expressed annoyance and established measures of personality such as neuroticism, or stability, and relationships between noise sources and respondent's source of employment and income do not, in general, influence noise annoyance. Furthermore, although complaints about a noise source often increase dramatically during media exposure or publicity campaigns, noise annoyance, as measured on noise annoyance scales, persists for sources which are not the subject of public attention (e.g., railway noise [5]). Possible bias introduced by drawing attention to a noise source can be eliminated by employing 'concealed' questionnaires which are ostensibly about other environmental matters and only introduce noise unobtrusively. There is evidence to show that even this form of possible bias might not be important, as surveys which included up to thirty minutes' discussion on noise have not shown any shift in expressed annoyance between concealed items at the beginning and repeated, open items at the end.

A consistent picture has emerged from a number of studies of the nature of expressed noise annoyance. At high noise levels, it is correlated with various forms of activity disturbance and communication interference. Annoyance reactions are not only affected by the level of a particular noise source, but also by the degree to which the individual holds certain attitudes, and the individual's degree of noise sensitivity. Greater expressed annoyance is associated with the belief that the noise could be prevented (misfeasance) and that health may be affected by noise. In a recent field study of noise annoyance associated with a regular helicopter shuttle service it was noted that certain individuals expressing high annoyance held the incorrect belief that passengers could look down on them. Thus, the sound of the helicopter became annoying, partly because it reminded those individuals of their belief in the invasion of their privacy. Fear of airplane crashes is strongly associated with expressed annoyance to the noise of airplanes flying over.

Notwithstanding the various attitudinal and psychosociological variables which can affect perceived environmental quality as expressed in terms of noise annoyance, the correlation with acoustical factors alone is generally sufficiently high (when considering the mean scores of homogeneous groups of persons) to enable perceived environmental quality to be used in selecting environmental noise criteria. Furthermore, the quantification of the subjective effects of the various non-acoustical variables constitutes a fruitful research topic, with obvious applications in terms of noise criteria. For example, a recent comparison of annoyance reactions to railway noise versus aircraft and road traffic noise has shown that railway noise is less annoying at the same noise level. This difference is presumably partly due to differences in attitude to railway noise and other transport noises (a non-acoustical factor), as well as acoustical factors, and could be included within general forms of noise criteria, or allowed for in source-specific noise criteria.

3. THE IDEAL

We can now take this opportunity to summarise the factors involved in the selection of an ideal noise criterion. First of all, the criterion must be consistent with the mean annoyance response of the exposed population. Secondly, the noise scale or index must be capable of being measured reliably, accurately and repeatedly. Finally, the criterion must be used and be seen to be reasonable and effective, and then it may become more and more widely used.

Unfortunately, because of the known large individual differences in expressed annoyance scores at each noise exposure level, it is not possible to devise any noise criterion that will work equally well for all individuals (see Figure 2.3). Thus, the mean annoyance response must suffice, although, where the costs of noise control permit it, it is possible to choose an annoyance response percentile above the mean in order to protect a proportion of those persons with above average sensitivity. Conversely, when costs would be excessive, it may be necessary to choose a percentile below the mean.

A further problem arises with non-acoustical variables. It is a known fact that the relationship between expressed annoyance and noise level is source-specific (as mentioned above). No simple noise scale differentiates between sources in any way that could be used for the selection of criteria. Thus, one of the goals of the Noise Advisory Council (NAC), to develop a single unified noise index for the selecting of criteria for all noise sources, remains some way from solution. It is perfectly possible, and even reasonable, to measure all noise sources in terms of a common noise index, for example, L_{eq}, but it is a different matter entirely to select levels of L_{eq} that can be used to define criteria independently of source. L_{eq} is a sensible unified noise index for use in complex noise environments for comparing and contrasting the energy contributions of each source. It can also be used to show how the energy contribution of a single source can change in importance with future development. However, it cannot be used in isolation to predict

Figure 2.3 Individuals' annoyance score against railway noise level [14]

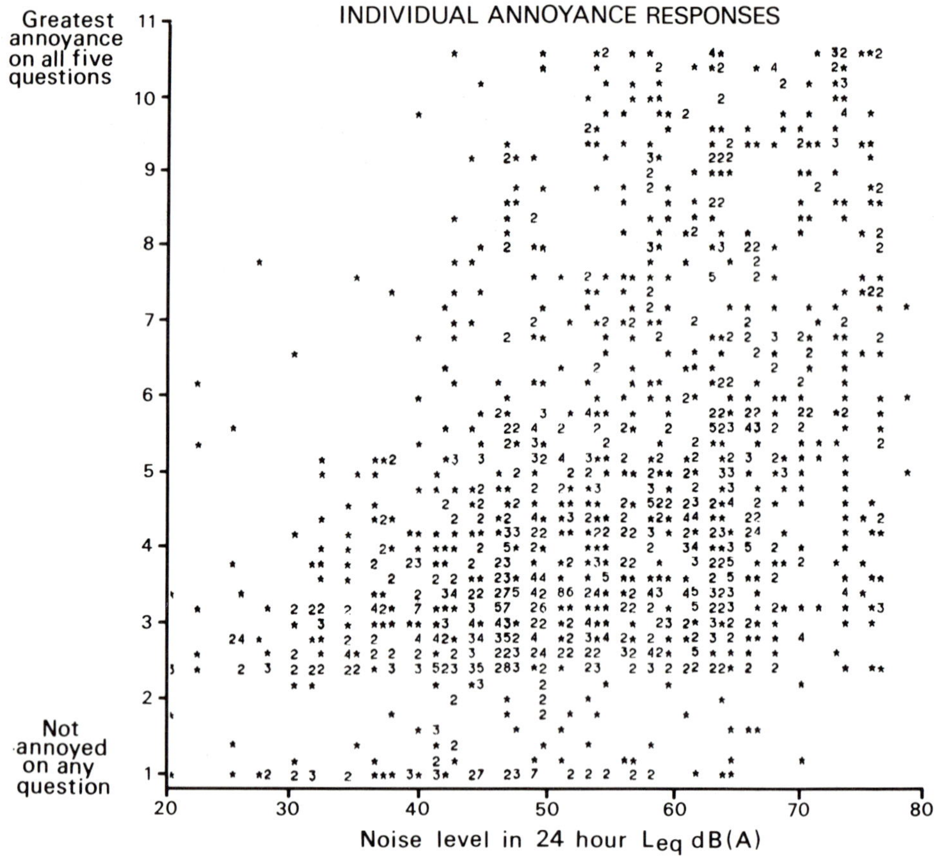

reliably the contributions to overall expressed annoyance from each source, as it is possible that any one source could be dominating the mean annoyance response, even though it has only a similar L_{eq} level to the other sources.

There is some evidence from laboratory studies on multiple noise sources that, even if the effect of different dose-response relationships for different noise sources is weighted out, a combination of two sources at equal subjective levels gives rise to much more annoyance than would be predicted on the basis of energy summation alone [6]. Thus, the effect of the introduction of a new source at the same L_{eq} level as the existing noise environment could be compared with the effect of increasing the existing noise level alone by up to 10 dB, instead of the 3 dB predicted by a simple L_{eq} model.

The problem of weighting for different sources is no more simple than the problem of weighting for different times of day. Although it is true that individuals may be more sensitive to noise during particular periods of a 24 hour day (for example, when trying to go to sleep), research to date has been inconclusive as to the amount by which noise levels should be weighted during the sensitive periods. In part this may be due to the use of data which have only been obtained indirectly: for example, in the Second Heathrow Aircraft Noise Survey [7] it was noted that scores on a night-time annoyance question (asked during the daytime) correlated less well with night-time exposure than did general annoyance with daytime exposure. Furthermore, general annoyance correlated more highly with night-time exposure than did specific night-time annoyance. This may have been because daytime exposure is more important in determining general annoyance reactions and that the correlation between night and day exposure tends to be stronger than any tendency for expressed night-time annoyance to correlate with night-time exposure. The technique adopted at present by Vernet in France [8] may have greater face validity. Disturbance of sleep by transport noise has been studied through in-situ physiological recordings. However, this approach suffers from having no direct transformation to perceived environmental quality, as directly measured sleep disturbance has not yet been related to expressed annoyance. A collaborative study has been planned in which laboratory annoyance measures (using ISVR techniques) will be related to M Vernet's sleep disturbance measurement technique.

Thus, it can be seen that there is no conclusive evidence for the adoption of any particular night-time noise weighting as, for example, 10 dB in the American L_{dn} index. Recent laboratory work, again using indirect questioning techniques, suggests that a 5 dB penalty could be more appropriate for night-time noise. There is some evidence that, given the way in which noise levels typically drop dramatically at night-time, the evening period may be the most important as far as annoyance responses are concerned. A recent combined field and laboratory study at ISVR showed a high correlation between field and laboratory scores when field questionnaires and the laboratory phase were closely related to the 19.00 and 21.00 hours evening period.

In fact, there is much to be said in favour of combined laboratory and field studies of the annoyance reaction to various noise levels. The laboratory technique offers an opportunity to present experimental subjects with a wide range of closely controlled conditions and, if it is closely tied to a social survey, it can be validated against responses to real-life long-term exposures.

In summary, it can be stated that there is, at present, no fully validated unified index which could be used to replace the variety of separate indices for specific sources which will be described in the next section.

4. CURRENT NOISE INDICES

This section will give a brief description of the noise indices currently employed for the setting of criteria in the United Kingdom, in the light of the points covered above.

For a significant number of persons living near to major airports the noise caused by jet aircraft flying over can constitute the major contributor to the noise energy received at their dwellings. A social survey and noise measurement programme was carried out in summer 1961, involving 1909 interviews and measurements at 85 points within ten miles of London Heathrow Airport. The Noise and Number Index (NNI) was developed from these data, and was based on the observation that multiplying the number of aircraft by four, without change of noise level, was equivalent to raising the noise level of each aircraft by 9 PNdB, without change in number, in terms of expressed annoyance. (PNdB, perceived noise level in decibels, is a complex form of frequency weighting of a noise and was developed from laboratory studies of aircraft flying over. Generally speaking, for typical aircraft fly-overs, the perceived noise level is approximately 13 dB greater than the 'A' weighted sound-pressure level.)

The NNI was adopted by the Government for the assessment of noise around major airports, and contours enclosing areas having specified NNI levels (or above) are regularly produced. These contours are produced by computer programmes and represent air traffic on a busy 12 hour summer day. The contours are then used to select criteria for acceptability. For example, the Greater London Council (GLC) guidelines suggest that the area within the 45 NNI contour is suitable for new dwellings only as infilling in established residential areas. The area within the 60 NNI contour is considered definitely unsuitable for any dwellings. Furthermore, it is possible to select a particular contour, e.g., 50 NNI, for the provision of noise-insulation grants, bearing in mind the costs of insulating large numbers of dwellings.

The particular dose-response relationship implied in the NNI has been examined in a repeat survey at Heathrow in 1967 and in a number of studies at, or near, other airports. Although no compelling reason has emerged for changing the form of the NNI index, recent laboratory studies [9] have indicated that the number versus level trading relationship is more complex than the simple relationship implied in the 15 logN term in the NNI.

Furthermore, it is open to question whether the NNI is applicable to a green-field site, although it may well correlate with response as well as any other known index. It is an index that is not intended to take account of the levels of other sources, which may or may not be important, and it is plainly not intended in any way to represent the dose-response relationships to sources other than aircraft noise.

Road traffic noise is currently assessed by using the L_{10} index; that is, the A-weighted sound-pressure level exceeded for 10 per cent of the time. The Noise Insulation Regulations 1973 [10] specify a criterion level of 68 dB(A) L_{10} (18 hour), which is the arithmetic mean of the 18 hourly L_{10} levels covering the period 06.00 to 24.00. The L_{10} index was selected as a simpler alternative to the Traffic Noise Index (TNI) proposed by Griffiths and Langdon [11], after a survey of 1400 residents of dwellings in London exposed to traffic noise, and it was adopted at a time when simple machines for measuring L_{eq} directly were not available.

The L_{10} index suffers from the inability to take account of either the peaks in individual events above the L_{10} level, or the degree to which the steady background drops below the L_{10} level. It is possible to envisage situations which, although subjectively different, could have similar L_{10} levels. As such, it is an inappropriate index for use with situations other than road traffic, and may not be suitable for situations where there is another noise source of comparable or higher levels than the road-traffic noise.

The GLC guidelines take a different approach with regard to road-traffic noise and specify indoor noise levels to be aimed at as acceptable criteria. Thus the architect is encouraged to design for a chosen indoor environment, irrespective of the outdoor noise levels. These guidelines are an internal L_{10} of 50 dB(A) in the daytime (07.00 to 19.00 hours) in living rooms, and 35 dB(A) in the evening (22.00 to 24.00 hours) in bedrooms. However, it is recognised that it may not always be possible to achieve these levels.

Industrial noise is commonly assessed in terms of BS 4142: 1967 (amended 1975) [12] which is a procedure for comparing a corrected noise level (CNL) from the offending source with the background level. The difference is used to predict the likelihood of complaints, on the premise that the greater the degree of intrusion above the background level the greater the likelihood of complaints. The procedure is straightforward and easy to apply, and is a favourite among local authorities, who often attempt to apply it outside its brief (referring specifically to noise from factories, industrial premises and to other fixed installations).

As a procedure, BS 4142 is limited in application and may be inappropriate in complex environments including significant contributions from transport noise. For example, were it to be erroneously applied to aircraft noise, it is likely that in many situations it would predict highly adverse community reactions where no such reactions occur and where the NNI levels are comparatively low.

The American Preferred Noise Criteria (PNC) curves or the European Noise Rating (NR) curves are often used as an alternative to the dB(A) scale for evaluating indoor environments. The PNC and NR values are clearly specified in terms of appropriate indoor use at different levels and are useful for evaluating not only indoor dwelling noise levels but also industrial, office and other more or less sensitive premises. For example, it is suggested that broadcasting studios should not exceed the 20 PNC curve (approximately 30 dB(A)) or the 15 NR curve, whereas private offices may have internal noise levels up to the 30 to 40 PNC curves or the 40 NR curve.

However, the above two indoor criteria systems, again, may not adequately allow for the effects of a mixed noise environment where there may be a number of competing sources. In general, they are very useful for assistance in the design of ventilation and similar plant, and indoor/outdoor noise attenuation, but they do not deal with perceived environmental quality and are of limited use in the control of outdoor levels from transport noise.

Finally, we come to the Equivalent Continuous Noise Level (L_{eq}) [15], an index which has a great deal of merit from the point of view of objective measurement, but which nevertheless should be used with caution for predicting community response. The L_{eq} unit is as fundamental to environmental noise as the kilowatt is to electric power but, taking the same analogy, just as the wattage of an appliance alone is not a sufficient definition of that appliance, neither does the L_{eq} alone sufficiently define any specific noise environment.

However, the L_{eq} index has been used to select criteria for noise from railway traffic, contruction sites and noise abatement zones. The GLC guidelines suggest some form of action where railway noise exceeds 65 dB(A) in terms of 24 hour average L_{eq} (for example), and is supported in terms of using L_{eq} by a recent national survey of railway-noise annoyance. This survey found that the 24 hour L_{eq} index is certainly as highly correlated with expressed annoyance as any other objective noise measure (see [14]).

5. CONCLUSIONS

The current UK practice of defining environmental noise criteria in terms of perceived environmental quality is probably the best way in which the problem could be tackled. The alternatives of using more objectively measurable human effects for the selection of criteria do not have as much appeal, given the present state of knowledge. However, this is not to say that the situation might not change if some method is found for relating such measures as, for example, physiologically measured sleep stage changes or long-term epidemiological effects to perceived environmental quality. Of course, such changes would require more research.

However, the current UK practice of separate indices for each offending noise source is not an ideal solution to the problem. Unfortunately, no completely comprehensive unified index has yet been discovered, despite the early promise of L_{eq} and its derivative, the Noise Pollution level [13]. The best policy, at the moment, is to continue to use the existing UK indices within the framework of their legally defined applicability, although it must be admitted that any decision to apply these indices outside the range of conditions which were included in their development is administratively convenient but not scientifically proven. Where 'difficult' cases arise, for example, the complex noise environment existing around a major airport and including road traffic noise, railway noise, industrial noise and various forms of airport ground noise, as well as airborne aircraft noise, it is politic to include an objective assessment in terms of the L_{eq} contributions from each source. Then the different noise sources can be compared and contrasted in the same unit, something which is very difficult when different indices are in use for each source. The overall assessment by the noise expert must then, in the present state of knowledge, include an element of professional subjectivity, taking into consideration not only separate assessments of each source considered alone in terms of its established index but also the L_{eq}-based contribution breakdown described above.

Where assessments must be made in respect of the complex situation outlined above, in terms of particularly sensitive aspects such as night-time noise, then reference must be made to concepts such as a consideration of a comparison of the degree of intrusion of offending noise events above the background noise level. Such an assessment would be complementary to whatever criteria may have been laid down in any particular case with respect to any single source considered alone.

Of course, not all situations are as complex as the example given, but it is as well to be aware of the capabilities and limitations of any tools that are available. It is to be hoped that further research, such as is being conducted at ISVR and other institutions, may shed some light on some of the difficulties referred to in previous sections of this chapter.

REFERENCES

1. British Standards Institution. BS 4197:1967. *A precision sound level meter* (1967).
2. Large, J.B., Rice, C.G., Walker, J.G. and Fields, J.M. 'Noise research and criteria', Paper presented at the Noise Advisory Council Seminar, Darlington, Sept 5–7 1978.
3. Bastenier, H., Klosterkoetter, W. and Large, J.B. *Damage and annoyance caused by noise*, Environmental Quality of Life, Commission of the European Communities, Report EUR 5398e (1975).
4. Miller, J.D. *Effects of noise on people*, US Environmental Protection Agency NTID 300.7 (1971).
5. Fields, J.M. and Walker, J.G. *Reactions to railway noise: a survey near railway lines in Great Britain*, ISVR Technical Report No. 102, Institute of Sound and Vibration Research, University of Southampton (1980).
6. Powell, C.A. *A summation and inhibition model of annoyance response to multiple community noise sources*, NASA Technical Paper 1479 (1979).
7. Office of Population Censuses and Surveys. *Second survey of aircraft noise annoyance around London (Heathrow) Airport*, HMSO, London (1971).
8. Vernet, M. 'Effect of train noise on sleep for people living in houses bordering the

railway line', *J. Sound Vib.*, **66**(3), 483–92 (1979).
9. Rice, C.G. 'Investigation of the trade-off effects of aircraft noise and number', *J. Sound Vib.*, **52**(3), 325–44 (1977).
10. Noise Insulation Regulations 1973. Statutory Instruments 1973, No. 1363.
11. Griffiths, I.D. and Langdon, F.J. 'Subjective response to road traffic noise', *J. Sound Vib.*, **8**(1), 16–32 (1968).
12. British Standards Institution. BS 4142: 1967 Amended 1975. *Method of rating industrial noise affecting mixed residential and industrial areas.*
13. Robinson, D.W. 'Towards a unified system of noise assessment', *J. Sound Vib.*, **14**(3), 279–98 (1971).
14. Fields, J.M. and Walker, J.G. Unpublished data (1979).
15. Noise Advisory Council. *A guide to measurement and prediction of equivalent continuous sound level* L_{eq}, HMSO, London (1978).

3 Experience of noise problems

Rupert Taylor

Noise is a subject of concern in building design for several distinct reasons. In each case the standard of design is dependent upon whether the source is external or internal, whether it is primarily from aircraft, road traffic, industry, construction or other sources, and whether it is caused by mechanical services or is due to human activity. Whatever the circumstances, there are always unforeseen pitfalls for the architect and engineer, most of which could be avoided at the outset.

1. SOUND INSULATION

As in many areas of acoustics, the non-acoustician is often under some important misconceptions about the fundamentals involved. Broadly speaking, before any structure can offer sound insulation of n decibels, it must occlude the area concerned so as to leave gaps of no more than $10^{-n/10}$ of the total area to give its full value, and, if it is only, say, 99.9 per cent effective, it cannot give more than 30 dB insulation, except at long wavelengths relative to the size of the gap. This is the most common error in the design of buildings, particularly where the openings and gaps are unsuspected. For instance, industrial buildings often have a row of openings the full length of the eaves when the roof is of corrugated sheet, and, on a smaller scale, few door designs, even of acoustic doors, have seals which are 100 per cent efficient, and often substantial sums are spent on heavy, multi-skin doors which give only 30–35 dB sound reduction because of poor seals. Although openings are clearly of the greatest importance, weak areas which may be impervious, but of low sound reduction, are often overlooked. Their effect is not so serious, however, and most textbooks publish a chart for estimating the composite insulation of combinations of areas of different sound insulation.

An example that shows the problems which result from failure to grasp the importance of small openings was a case where a publishing company intended to acquire the lease of a new office block under construction close to a busy main-line railway. Mindful of the problems that could be caused by the 99 dB(A)-plus peak external sound levels from passing trains, the prospective tenant pressed the landlord on the subject, and the latter supplied a consultant's report on the required sound reduction of the windows and an assurance that the windows to be supplied would meet the specified standard. However, the window specification was adequate only when all windows were closed, and, although the building contained mechanical ventilation, no cooling was provided, and the probability that some windows would be opened in summer was very high. The effect of opening windows of an area of three square metres would range from 66 dB(A) in the open-plan areas compared with 57 dB(A) with windows shut, to a figure of 96 dB(A) throughout a partitioned office.

Provided that the noise is primarily escaping through the structure, the design task is one of applying well-established acoustical principles, and indeed most normal building specifications provide adequate sound reduction, with the exception of lightweight cladding and, most important of all, party walls between dwellings. This is an increasing problem as standards of workmanship decline. Bad bricklaying, causing porous walls, or the laying of bricks frog-downwards without care to fill the frog with mortar, are common problems, as is the bridging of cavities with mortar. There is still some confusion among professional people about the different roles of sound-absorbent and sound-insulating materials. This confusion manifests itself, for instance, in the frequent cases where architects specify suspended acoustic ceilings in a building where demountable partitioning, selected for a high sound-reduction index to achieve speech privacy, is to be installed. The partitioning then extends only from floor to suspended ceiling and, as the sound-reduction index of many acoustic ceilings is often only 15 dB or so at speech frequencies, the transmission of sound via the ceiling void limits the overall sound insulation of the partitions to a figure often 10–15 dB below its quoted performance. There are two solutions to this problem. The best is to specify impervious suspended ceilings, the sound transmission loss of which in this situation is as great as that of the partition. This is achieved for most partitions if the ceiling incorporates, or is inherently equivalent to, a sheet of 9 mm plasterboard as a backing to the sound-absorbent medium. This solution is valid regardless of the present or future location of the partition. The second solution involves the construction of an impervious bulkhead to seal off the void above the partition, either of plasterboard or of a proprietary flexible heavy curtain or lead sheet. However, apart from the difficulties of sealing such a barrier around ducts and other obstructions, there is also the drawback that it severely limits the flexibility of the partition layout.

A major problem in achieving high sound insulation is a lack of understanding on the part of the workmen of the reasons for what they are doing, if it represents any departure from normal practice. Many designs rely on the use of independent structures, such as separate stud frames for the two leaves of a plasterboard partition, but the chances of avoiding unintended bridges are very low unless there is frequent inspection of the work as it proceeds. Drawings are one thing; the result of builders' initiative is another. The sound insulation of partitions, even with correct ceiling-void treatment, often falls down because of poor detail at window mullions, or where pipes or ducts run along the structural wall and the only consideration given to acoustical isolation is to pack a little glass fibre inside a pipe casing.

Windows, even when sealed, present their own particular problems. Every designer knows that 'double glazing' is proof against noise, and many are even quite conversant with the rules concerning the depth of air cavity and the thickness of the panes. The difficulties arise when, for instance, the sound source against which insulation is required is heavy road traffic. Windows with large dimensions often have panel resonances which coincide well with the frequencies of peak emission from the exhaust systems of heavy lorries, and double windows of large dimensions have similarly low-frequency cavity resonances. Window specifications in buildings to be subjected to aircraft noise may include extra thick glass, notwithstanding that its thickness causes coincidence effects at frequencies close to the most significant region of the frequency spectrum of aircraft noise.

2. ROOM ACOUSTICS

While problems of sound transmission from inside to outside, and vice versa, are largely those of sound insulation, the influence of room acoustics can never be overlooked, and of course problems involving source and receiver located within the same room or building can be strongly influenced by the acoustics of the building. Often the effect is quite misleading to the layman, owing to the indirectness of the link between reverberation time and reverberant sound level. A shipbuilding company had a joinery shop located in one of the aircraft-hangar-like buildings common in shipyards, and over many years tolerated the noise from the woodworking machines, despite the reverberant nature of the cavernous building. When the joinery shop was rehoused in a bright, new, modern partitioned area, with a ceiling height of only three metres and a plan area a fraction of that of the building as a whole, they were quite perplexed to find that in the new room, which had a much shorter reverberation time than the old building, the reverberant-sound level was substantially higher. The reason is that reverberant-sound level is an inverse function of the surface area of the room and its average sound absorption coefficient (and for the same surface characteristics, of the surface area alone), whereas reverberation time, for the same surface characteristics, is a function of the room volume.

Another deceptive feature associated with sound-absorbent materials, such as commercial acoustic tiles, is that they are often quite significant reflectors, particularly at grazing incidence. First of all, a fibre acoustic tile, say, with an absorption coefficient of 0.7, will only reduce the intensity of singly relected sound by $10 \log (1 - 0.7) = 5.2$ dB. Furthermore, quoted sound-absorption coefficients are average figures for all angles of incidence, whereas the absorption coefficient at grazing incidence may be quite different, particularly where there is a perforated impervious facing.

This effect is important in, for example, open-plan offices, particularly where acoustic screens are used, and an important component of the received sound is that reflected off the ceiling. There are significant advantages to be gained from using ceiling designs which are known to have high absorption coefficients at low angles of incidence in these situations.

Where the noise is due to human activity, such as in a situation common in the City of London where large numbers of dealers in currency or commodities sit at consoles and spend most of their time having telephone conversations with buyers and sellers, the acoustics of the room play a double role. Not only do they influence the level of background noise, but, since high levels of background noise reduce the intelligibility of speech, the levels of speech are raised, which in turn raises the level of background noise, and so on in a vicious circle. Increasing the sound absorption in the area not only reduces the reverberant sound level but also results, very often, in lower voice levels, thus reducing the background level by substantially more than the predicted effect of the increased absorption alone.

3. MECHANICAL SERVICES

The control of noise from mechanical services is a somewhat specialist area of acoustics. Clearly, the acoustic properties of the building served have an important role to play in determining the final noise levels, but there are also other important factors as well, notably the transmission and generation of sound in ducts, and the structure-borne transmission of sound in buildings and in ducts and pipes.

Modern building structures are highly efficient waveguides for propagating sound of a structure-borne nature and it is consequently of the utmost importance to control the transmission of sound energy from plant and machinery directly into the structure. The basic stipulations concerning provision of vibration isolators and flexible connectors have become part of standard engineering practice, and the problems which arise are seldom due to failure to observe fundamental rules, but to unforeseen errors. A common problem arises with large refrigeration compressors. The developers of a shop and office building installed a reciprocating compressor in a loading bay, mounted on suitably soft springs, and fitted with braided flexible metal bellows-type connectors. However, because of the high pressures in the refrigerant and the stiff characteristics of the connectors, very little attenuation of the sound power generated by the compressor was achieved. Since the compressor was in a position remote from the evaporator and condenser, compressor noise was radiated structurally through the pipe hangers in all areas of the building through which the refrigerant pipes passed. The only effective solution was to provide flexible mountings for the pipe hangers, because of the extreme difficulty of achieving greater flexibility in connectors for high-pressure refrigerant and the expense of fitting them as an afterthought. This type of problem is not always foreseen by the design engineer, who sees flexible connections on the drawing and is satisfied.

Sometimes other isolation devices turn out, in practice, to be not quite what they seem. Isolators on fan units can have top bolts screwed right down to the floor slab. Flexible duct connectors can cover over a duct misalignment where the two ducts touch. Floating slabs can be concreted up at the edges. Flexible pipe connectors can be anchored on the wrong side.

One of the most awkward problems concerns the question of fan bearings. In a centrifugal fan, the main impeller bearing is often rigidly bolted to a steel pedestal fixed to the scroll, or directly to the scroll itself. Consequently, the whole fan acts as an efficient impedance-matching device for feeding bearing noise into the airstream. This is taken account of in the manufacturers' quoted sound-power levels, but it only applies to bearings in good condition. As soon as the bearings begin to wear, a characteristic narrow-band noise signal is injected into the normal fan noise and in critical cases appears in the area served by the system. The difficulty is caused by the fact that bearings which are noisy often function quite well, and would not normally be replaced as part of routine maintenance.

Lack of maintenance in other areas can cause some very odd effects. An old system for extracting air from the basement of a city office block involved some moderately large centrifugal fans mounted in a roof plant-room. Latterly, a new tenant of the building developed the area beneath the plant-room as a boardroom, and was disconcerted to discover that there was a high level of structure-borne noise in the new boardroom. Inspection of the plant-room above showed that there were no vibration isolators, and the fan and motor were bolted directly to the floor slab. To install isolators retrospectively would incur large expense because of the need to construct steel subframes for the fan and motor assemblies and to realign the ducting. There was, however, something uncharacteristic about the noise which prompted closer investigation. There was a curious impression given that the noise originated from some secondary source, located in the discharge ducting, rather than from the fan itself. Inspection of the duct revealed that the discharge grille was totally blocked with greasy soot from extracting thirty years' worth of London air. No air was any longer being extracted and the fan was working against a closed outlet. Cleaning of the grille immediately solved the problem, despite the lack of vibration isolators.

There is an unfortunate tendency among designers of buildings to think that the top of the building is a good place both for plant-rooms and for boardrooms, not to say directors' flats. There is one office block which has a small power station containing three 800 kW diesel generators on the eighth floor, and the architect specified a noise level of Noise Criterion 25 for the seventh floor, which was to have been used for directors' offices and a boardroom. In the event, they had to be content with NC 35, and the area was used as a staff restaurant, but even that required a level difference of 70 dB(A) between the sound level in the engine room and that in the area below. Needless to say, it was not achieved without considerable expense, out of proportion to the need to place a boardroom directly beneath the generator room. This particular job highlighted another possible cause of error on the part of design engineers. It was, of course, necessary to install the generator sets on highly efficient vibration isolation systems, which involved a mass-spring-mass-spring system of concrete inertia block on isolating pads, supporting the generator sets on unit isolators. The isolator design engineer calculated the transmissibility as the product of the transmissibility of two independent mass-spring systems, instead of working out the couple frequencies of a mass-spring-mass-spring system. It also drew attention to the lack of data on the transmissibility of rubber-type isolators at acoustic frequencies. Despite these problems, the result was acoustically impressive.

Duct systems play an important part in determining the overall efficiency of sound-insulating structures. The design engineer is often quick to notice the need for cross-talk attenuators where a single-duct system has openings into two areas requiring a high degree of mutual separation, but overlooks less obvious problems. Long runs of duct through suspended ceiling voids can have a great enough surface area to offset completely the sound-reduction index of the duct wall, and they function as highly efficient aerials to collect noise from one area and radiate it into another. When the duct is rectangular and thus of low stiffness, this can cause problems at low frequencies. Attenuators in ventilation systems are not usually needed to provide high insertion losses at low frequencies, because the loss due to end reflection at the duct termination is normally enough to leave an attenuation requirement only at frequencies above the 125 Hz octave band. However, a long run of rectangular ducting in a suspended ceiling void effectively emits all its low-frequency noise through the duct wall before it reaches the duct termination and, since the ceiling will offer little resistance to low-frequency noise, an unpleasant situation with high levels in the 31.5 Hz octave band results. There is little that can be done to cure this problem retrospectively.

Low-frequency noise from a number of sources can be an unexpected cause of problems for two reasons: the sound insulation of most structures is at its lowest at the bottom end of the spectrum, and the frequencies at which standing waves are easily set up in most rooms are similarly low. On top of this, the increasing tendency to measure environmental noise in units of 'A' weighted decibels leads to an underestimate of the significance of low-frequency noise, because of the steep slope of the 'A' weighting curve at low frequencies, which was originally intended for sounds of low intensity only. The almost universal tendency towards the use of gas- and oil-fired burners in heating systems has introduced a new and widespread source of noise in the 31.5 Hz octave band (which may go some way to explaining the existence of so-called 'hum' phenomena). Solutions to low-frequency problems require specialised knowledge.

4. NOISE FROM AMPLIFIED MUSIC

It is indeed the high-power output of modern music reproduction systems at low frequencies which is largely responsible for the fast growth of one of the biggest problems associated with living in flats or semi-detached and terraced houses. Building standards were devised long before the 20 watts per channel stereo amplifier became almost as common as the television and, even if they had not been, there is no way in which party-wall specifications could practicably be made proof against such an onslaught. Low-frequency sound insulation is stiffness-controlled rather than mass-controlled, but increased stiffness means lower sound insulation in the mid-frequency range. It is, regrettably, difficult enough to get builders to achieve traditional standards of sound insulation in party-walls.

Practitioners in building acoustics receive frequent enquiries from members of the public about methods of improving the sound insulation of walls, floors and ceilings, and, if there were a simple 'stick-on' product which would have a worthwhile effect, its inventor would make himself a fortune. Unfortunately, there is no escaping the mechanical properties of air, which means that add-on structures must have either extremely heavy skins or very deep airspaces to avoid having natural frequencies right in the middle of the band of frequencies where the improved sound insulation is required. Building Research Station have a data sheet which describes the best compromise, which consists of a plasterboard false wall on a timber stud frame, supported at the edges only, and with a quilt in the cavity. Variations on this theme are possible for ceilings as well but it soon becomes an expensive and difficult job. Many of the problems occur in conversions of large town houses into flats, and unless the house was exceptionally well built in the first place (and some Victorian builders knew far more about sound insulation than most of those of today), or has had its floor/ceiling specification suitably improved, intolerable conditions can result.

5. CONCLUSION

Acoustics will always be a specialist subject, and non-specialists may sometimes make mistakes – sometimes of an expensive nature. Fortunately, the number of specialists is increasing, and, as it does so, their advice will become more readily available. The biggest problem to be overcome is the tendency among members of other professions to feel that they ought to be able to cope with acoustical questions and that it is something of a defeat to have to turn to another profession for help.

4 Noise sources in buildings

Tim Redmore

SYNOPSIS

The aim of this chapter is to describe both the internal and external sources of noise that can affect the sound level in a building. Methods of predicting the sound level that a source produces, together with the associated noise spectra, are included where possible. Relevant aspects of design, installation and maintenance of equipment are discussed, as these can affect the resultant sound output. Emphasis is placed upon co-operation between the individual members of the design teams, so that they work together for a good overall plan and not just for their own individual section. The economic benefits of correct planning, as opposed to the cost of post-construction remedies, are included. In order that noise sources can be grouped into sections, noise is considered to originate from either outdoors, mechanical engineering services, electrical services or people.

1. INTRODUCTION

The noise levels within a building, and the acoustics of the spaces within the building, are as important as other building services, such as heating, ventilation and lighting. An acceptable environment in which to live or work cannot be achieved without due consideration of these acoustic factors. Because noise within buildings is produced by a wide variety of services, the control of the noise is often left to the designers of these services, and an acoustic engineer is only involved if the finished product is not satisfactory. This method of planning should be good enough, provided that the designers understand the noise problems that might arise.

In a survey of service system faults at the University of Liverpool, Doig and Mulholland [1] found that 50 per cent of the buldings looked at had failures and 30 per cent had serious faults. Five causes of failure were described:

(a) Incorrect design procedure in ascertaining the acoustic requirements
(b) Incorrect design
(c) Lack of supervision during installation
(d) Irregular maintenance checks
(e) Lack of consideration of noise problems due to a tight budget.

They further concluded that more attention should be paid to the problems of service system noise, if an acceptable acoustic environment in a building is to be achieved.

Croome [2] surveyed lecture rooms in universities by means of a questionnaire and site appraisal. The questionnaire found that the acoustics of about a third of 120 lecture rooms at 16 universities were considered unsatisfactory by lecturers, building officers or students, while the site appraisal found two sources of noise:

(a) The basic nature of the air-flow system in ventilation ducts
(b) Bad planning or installation of service equipment

It has been claimed that the cost of remedying a faulty installation can be between 200 per cent and 2000 per cent more expensive than providing a correct design (Schneider [3]). It is important, therefore, that the service designers understand the areas in their designs that can lead to noise problems.

External noise from such sources as road traffic and aircraft, and internal noise, from people and the machines that they use, add to the noise within a building.

Thus, the design team as a whole should understand how the design, installation and maintenance of equipment, together with the basic plan of a building, can affect the acoustic environment.

This chapter aims to describe both the internal and external sources of noise that can affect the sound level in a building. Methods of predicting the sound level that a source produces, together with the associated noise spectra, are included where possible. Relevant aspects of design installation and maintenance of equipment are discussed. So that the noise sources can be grouped into areas from which they originate, the chapter is divided into four sections:

- *External noise:* road, rail and air traffic.
- *Mechanical engineering services:* the basic plant and supply system for heating, ventilating, airconditioning and plumbing services.
- *Electrical services:* lighting and lifts.
- *People:* cross-talk, the noise from machines they use, such as typewriters in offices, and movement.

2. EXTERNAL NOISE

Three common sources of noise in buildings originate from outside – road traffic, railways and aircraft. In 1963 a survey conducted for the Wilson Report [4] showed that, when at home, 36 per cent of the people questioned were disturbed by traffic noise, 9 per cent by aircraft and 5 per cent by railways. A Road Research Laboratory Report [5] estimated that between 19 and 46 per cent of the urban population in the United Kingdom lived on roads producing undesirable levels of traffic noise. The report also estimated that if the growth in road traffic continued at the same rate, the range of population exposed to these levels would rise to 61 per cent in 1980. Obviously, the problem of rail and aircraft noise is more localised, but will cause problems in some buildings.

Knowledge of the amount of noise from these sources is needed at the planning stage of a new building, as insulation from external noise can be costly for a completed structure. If it is estimated that the external noise will be excessive, the internal layout of a building can be changed in order to protect the more critical areas. For residential dwellings this may mean having the bathroom and kitchen facing a road or railway, with the bedrooms and living rooms at the rear of the house. Alternatively, the external walls facing the noise source may have few or no windows in them, as the sound insulation provided by a window is much less than that of a solid wall. In offices, it may be desirable to have communication areas, corridors, stairs and lifts, situated round the perimeter of the building, thus providing extra distance and extra structure for the noise to penetrate before reaching the critical areas. This form of planning, though, is not always popular with the occupants, as many people prefer to be able to see outside and not feel totally enclosed.

2.1 Traffic noise

Although the maximum sound level that a vehicle can make is governed by the Motor Vehicles (Construction and Use) Regulations [6], it is the sound level that the total traffic flow produces that is of interest to the architect. There have been several scales proposed to measure the annoyance of traffic noise, among them Perceived Noisiness Level (PNL), dB(A) with L_{10} values, Equivalent Continuous Sound Level (L_{eq}), Noise Pollution Level (LNP) and Traffic Noise Index.

It is the L_{10} value, the level in dB(A) that is exceeded for 10 per cent of the time, that is used in Britain as a guideline to which designers can refer, while in 1971 the International Organization for Standardization recommended the use of equivalent continuous sound level [7]. L_{eq} is a method of averaging the sound energy and can be considered as the sound level which would cause the same A level weighted sound energy to be received as that due to the actual sound over a period of time. For busy roads in a free flow situation L_{10} and L_{eq} levels are similar, but at situations where the sound is intermittent there is a difference.

The Wilson Report of 1963 proposed the L_{10} levels shown in Table 4.1 should not be exceeded inside dwellings.

Table 4.1

Situation	*Acceptable levels* *Day*	*Night*
Country areas	40	30
Suburban areas, away from main traffic routes	45	35
Busy urban areas	50	35

Acceptable sound levels in other buildings will often be stated by the client in the form of a Noise Rating Curve (NR) or Noise Criterion (NC). This level will depend on the activity in the building and will change from room to room.

Table 4.2 shows the noise climate found in [8], in situations where traffic noise predominates. The noise climate is defined as being bounded by the L_{10} and L_{90} noise levels.

Table 4.2 [8]

Group	*Location*	*Noise climate dB(A)* *Day*	*Night*
A	Arterial roads with many heavy vehicles and buses (kerbside)	80–68	68–50
B	(i) Major roads with heavy traffic and buses (ii) Side roads within 14–18 m of A or B group road	75–63	61–48
C	(i) Residential roads (ii) Side roads within 14–18 m of heavy traffic routes (iii) Courtyards of blocks of flats screened from direct view of heavy traffic	70–60	54–44
D	Residential roads with local traffic only	65–57	52–44
E	(i) Minor roads (ii) Gardens of houses with traffic routes more than 61 m distant	60–52	48–43
F	Parks, courtyards, gardens in residential areas well away from traffic routes	55–50	46–41
G	Places of few local noises and only very distant traffic noise	50–47	43–40

Clearly, many of these locations depend on good sound insulation of the exterior walls of dwellings to attain indoor night levels of 35 dB(A). Also, many of the sites exceed the 68 dB(A) L_{10} level specified for new or improved roads by the *Land Compensation Act of 1975*, HMSO, above which premises may obtain grants for sound insulation. The Act, however, does not allow for existing roads, which means that many people are exposed to excessive noise levels.

In a survey of several building and road situations, Redmore [9] found excessive noise levels in buildings situated on existing roads. The survey also highlighted how the following points should be considered when assessing the traffic noise problem and how to deal with it:

(a) The predominant noise sources: cars, buses, lorries, motor-cycles
(b) The type of traffic flow: free flowing, intermittent, braking or accelerating
(c) Planning: the positioning of rooms, reflections from other buildings etc.

One example of excessive noise found during the survey was that of one occupant, a student who was used to a quiet environment at home, but lived in a noisy situation while at university, needing to wear ear defenders in order to get to sleep at night.

Although traffic noise levels are not dangerous to the ears, Kryter [10] describes the dangers in the sociological, physiological and economic sense. One of the major problems described by Kryter is the effect of traffic noise on sleep. The effect on sleep is rather complex, due to the different stages of sleep, but Kryter states that:

> *Richter, for example, observed that a sleeping subject exhibited EEG and VCR (vasco-constrictive responses) reactions about every thirty seconds due to cars, trains and motor cycles passing the test room, even though the person slept quietly and upon awakening had no recollection of any disturbances. Richter views these responses as indicating that the vegetive system is withdrawn from the recovery and strength gathering process of sleep, and further believes this withdrawal is detrimental to normal health.*

If a designer needs to estimate the noise problem of a road, there are two methods. Either measurements can be made at the site, or, if traffic flow figures are known, a prediction can be made, using the DOE booklet *Calculation of road traffic noise*, HMSO (1975). The prediction method estimates a basic noise level dependent upon flow, speed, and constitution of the traffic, together with the gradient and type of road surface. Correction factors for distance, reflections and screening are then applied to produce the predicted noise level. Whichever method is used, an early assessment of the problem is essential, so that allowance can be made at the planning stage either for a change in the basic layout of the building, or for the use of a structure with more sound insulation.

2.2 Railways and aircraft

The intermittent nature of rail and aircraft noise means that an L_{10} value does not give a true indication of the nuisance. Reference to the relevant time-table can give an indication of the frequency of disturbance, but will not show other important variables. Diesel and electric trains produce different sound levels, while railway cuttings can reduce the amount of noise transmitted to the neighbouring area.

The sound level reaching a site from an aircraft will depend upon the height of the flight path, and the direction of the flight path, as aircraft noise is directional, as well as on the make and model. In order to take into account the high-frequency noise of a jet engine, a different unit, the perceived noise decibel (PNdB) is used. The Wilson Report considered that a Noise and Number Index (NNI) of 50–60 constituted an unreasonable exposure to aircraft noise, where the Noise and Number Index is defined by:

$$\text{NNI} = (\text{average peak noise level in PNdB}) + 15 \lg N - 80$$

where N is the number of aircraft exceeding 80 PNdB in the specified period.

Sound-level measurements will show if any precautions are needed at the design stage in order to reduce the noise penetration into the building.

2.3 Industrial noise

Industrial noise can affect both the employed persons and the surrounding area. The effect of noise on employed persons is covered by a Department of Employment Code of Practice, HMSO (1972). Although the Code does suggest a few methods of controlling noise exposure, it deals mainly with the maximum levels of noise that employed persons should be exposed to, and states that the control of noise should be dealt with by the building designer and equipment manufacturers.

The Code states noise limits of:

(a) Continuous exposure: 'If exposure is continued for 8 hours in any one day, and is a reasonably steady sound, the sound level should not exceed 90 dB(A).'

(b) Non-continuous exposure: 'If exposure is for a period other than 8 hours, or if the sound level is fluctuating, an equivalent continuous sound level (L_{eq}) may be calculated.'

The Code also describes methods of making measurements, the calculation of the equivalent continous sound level, acceptable ear protectors and the responsibility of management and employed persons.

The effect of industrial noise on the surrounding area is covered by British Standard BS 4142: 1967. The standard defines a method of determining whether a noise is likely to give rise to complaints, by means of comparing a Corrected Noise Level (CNL) with a measured or notional background noise level. The Corrected Noise Level can be calculated by measuring the noise outside the building where complaints have arisen and applying corrections for intermittency, duration, and tonal and impulsive character, as specified by the standard. The notional background level is used when there is difficulty in establishing a meaningful actual value. A recommended method of calculating the notional level is included in the standard.

Industrial noise can, therefore, present several problems to the building designer. Careful planning is required where noisy machinery is to be located. Acoustic material on the room surfaces may be needed to reduce the reverberant sound level, while acoustic screening may be needed round some machines. Consultation with machinery manufacturers should be sought at an early stage to ascertain the noise levels produced by their equipment. Although ear protectors may be provided, education of employees to the dangers of exposure to noise is often necessary to convince them of the need to wear them.

The possibility of complaints from occupiers of surrounding buildings may affect the design of the building structure. Walls of more mass may be used, while windows, air vents etc. may be located in walls away from likely noise-sensitive areas, or in the roof. Noise levels different from those of BS 4142 may be specified by the planning department of local councils, who might work on the basis of not increasing the background noise level. Early consultation here is useful, and might save unneeded expense.

2.4 Other external noise sources

There are other external sources of noise. Some, such as refuse collection and milk delivery, occur at regular intervals, while others, such as construction work and road repairs, might last for a length of time and then cease. It is unlikely that a designer would pay much attention to these unless specifically requested, as the cost of eliminating the intrusion of such sound would generally outweigh the benefits.

Construction work, however, can provide problems for the contractor using noisy equipment. Besides safeguarding the site workers from exposure to excessive noise, precautions may be necessary so that nearby buildings are not affected. The Department of the Environment has produced an advisory leaflet on *Noise control on building sites*, HMSO (1972). The leaflet describes excessive noise as that which makes conversation in the nearest building difficult with the windows shut. The leaflet further states that:

> 'Noise levels, between say 07.00 and 19.00 hours, outside the nearest window of the occupied room closest to the site boundary should not exceed:
> 70 decibels (dB(A)) in rural, suburban and urban areas away from the main road traffic and industrial noise
> 75 decibels (dB(A)) in urban areas near main roads and in heavy industrial areas.'

Included in methods by which noise from construction work can be controlled are:

(a) Reducing noise at the source: the leaflet includes a table of noise sources together with possible remedies.

(b) The use of acoustic screens.

(c) The siting of the plant: locating noisy plant as far as possible from the nearest buildings, or the use of partially constructed buildings as screens can reduce the problem of construction noise.

3. MECHANICAL SERVICES

The heating, ventilating and plumbing services contain many potential sources of noise in buildings. It is important, therefore, that designers should understand which components are liable to produce a problem, and methods of predicting the extent of the problem.

3.1 Fans

There are three sources of noise from fans. They are:

(a) *Rotational noise.* The action of the impeller blade on the surrounding air produces impulses in the air. As the impeller blade passes the same point, the air at that point receives an impulse, and a discrete frequency, the blade-passage frequency, is emitted, together with some harmonics. The blade-passage frequency, f_b, can be calculated by:

$$f_b = nN$$

where n = number of fan blades and N = fan speed (rps).

(b) *Vortex noise*. This is produced by the turbulent flow of air across the fan blades and has a random frequency distribution.

(c) *Mechanical noise*. This noise originates from the motors, bearings, belts etc. but should be controlled by good design and installation, including good vibration isolation. Poor maintenance will also lead to mechanical noise.

Although manufacturers should be contacted wherever possible for information on the noise emitted by their equipment, an indication of the sound-power level can be obtained using the following formulae:

$$
\begin{aligned}
L_W &= 67 + 10\,\lg M + 10\,\lg H \quad (\text{dB re } 10^{-12}\text{ W}) \\
&= 40 + 10\,\lg Q + 20\,\lg H \\
&= 95 + 20\,\lg M - 10\,\lg Q
\end{aligned}
$$

where Q = delivered volume (m^3/s)
H = fan static pressure (N/m^2)
M = rated fan motor power kW.

These values are based on work by Allen [11] and Beranek [12], and apply to fans operating at their optimum working point. Finkelstein [13] discusses the various mechanisms of noise production in fans and includes related experimental results together with those of other workers. Figure 4.1 shows how the sound-power level of a fan varied with

Figure 4.1 Fan efficiency and noise output [13]

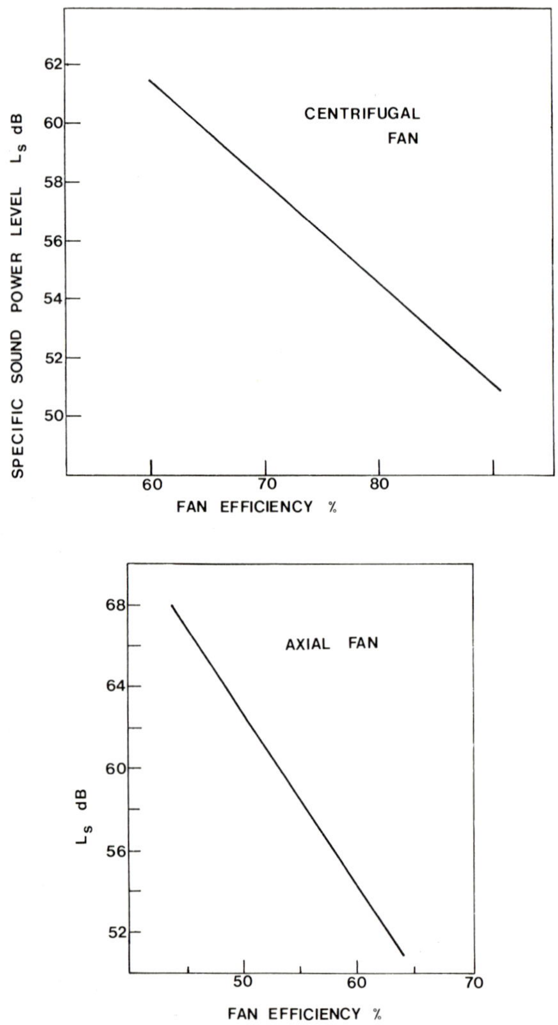

efficiency in Finkelstein's tests. These results agree with those of Bommes [14] in indicating that as the fan efficiency decreases, the specific sound level increases.

The spectrum of the sound depends on the type of fan. Typical spectra are shown in Figure 4.2.

It can be seen from the spectra that the noise from axial fans is mainly broad-band, while centrifugal fans are noisier at low frequencies but become approximately 5 dB per octave quieter at higher frequencies.

Figure 4.2 Typical fan-noise spectra [15]

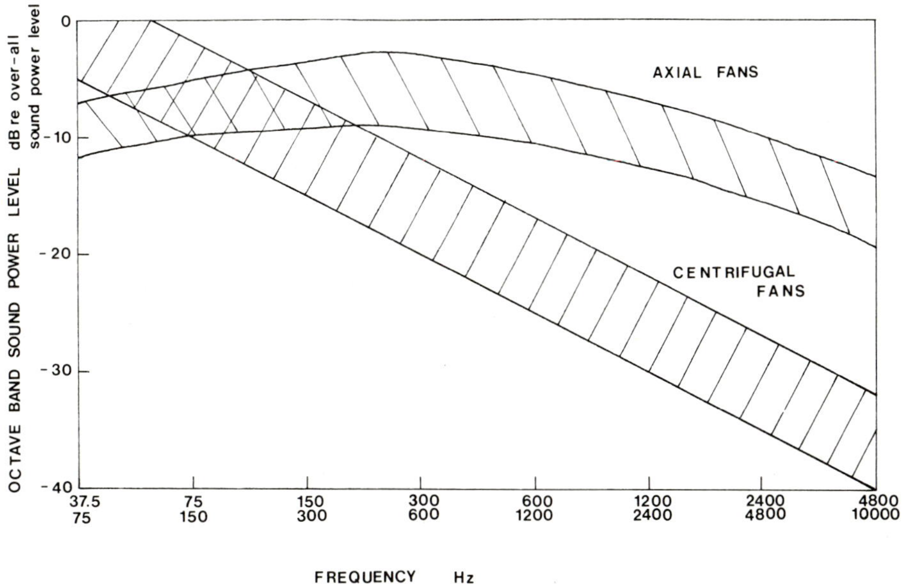

The shape of these spectra, together with any other attenuating equipment used in a system, can determine the choice of equipment. Most attenuators provide more attenuation at higher frequencies. Lower sound levels throughout the spectrum might therefore be obtained using an axial fan with an attenuator.

3.2 Duct noise

Although a sound is absorbed as it travels along a duct, especially if the duct is acoustically lined, the movement of air through ducts can lead to the generation of noise. There are several prediction methods available for the designer to calculate the sound transmitted through ventilation systems. These include the IHVE Guide, *Woods' guide to practical noise control* [16] and *The control of noise in ventilation systems* [17].

Until recently, it has been considered that, provided the air velocity does not exceed 10 m/sec, the sound produced by air movement, called regenerated noise, is negligible. Above this velocity, the action of the air flow encountering obstructions such as turning vanes, bends, grilles and even silencers produces turbulence which can lead to noise of a significant level compared with the overall system level.

The sound level produced at a fitting depends on the air velocity, the dimensions of the fitting, the turbulence created and the frequency being considered. Methods of estimating the sound level generated at fittings can be obtained from the IHVE Guide [15] and Atkins [17].

The IHVE Guide presents the following equation for predicting the sound-power level generated by a duct fitting, together with a table giving corrections for octave-band frequencies:

$$L_W = C + 10 \lg A + 60 \lg V$$

where C = constant that depends on the fitting and the flow turbulence
A = minimum flow area of fitting
V = maximum velocity in fitting (m/sec).

Badly designed ductwork with inadequate stiffening can also produce noise, usually a low rumble, if the surfaces are set in vibration by the flow of the air, or if they have not been isolated from other vibrating surfaces.

The continuing research into noise from ventilation systems indicates that existing methods are not as accurate as they might be. In a review of published literature up to 1974 concerned with service-systems noise in buildings, Gillam [18] lists nearly three hundred sources of information, covering items such as ducts, fans, pipework and refrigeration equipment. Gillam makes the point that, although research is aimed at a 'systems approach', much of the work has been compartmentalised. Thus, reference to many sources is required to obtain an overall picture of the problem of noise in buildings.

Doig [19] tested several duct fittings and compared the experimental results with traditional prediction data. The results showed:

(a) Mitred bends gave higher levels of attenuation than predicted, the difference being dependent upon the source.

(b) Branch ducts: a frequency dependence in the attenuation at a 90° tee-junction was found, although neither Sharland [16] nor the IHVE Guide method [15] give any difference across the spectrum. An example of this was that, for a 0.1 m x 0.1 m branch, a maximum attenuation of 8 dB in the 1000 Hz octave band was found, although Sharland predicts 12 dB and the IHVE Guide predicts 6 dB.

(c) Flow tests: normal attenuation did not occur if duct fittings were not several duct widths away from each other.

Doig concluded from his work that it is pointless to use prediction methods that 'work to the first decimal of dB's' since then the engineer 'adds a factor of safety that is of the order of 10 dB which eradicates the errors'.

In order to provide a fresh theory that accounted for the results obtained in his work, Doig developed a mode-coupling theory, applicable to mitred bends and tee-junctions.

Mode-coupling theory is not new, and a development of the theory can be found in Morse [20]. Morse shows that the pressure at a point in a room can be related to the source strength, the room modes at the source and the room modes at the point, by the following equation (ignoring time variations):

$$P(x, y, z) = U\Sigma[\psi(x_0, y_0, z_0)\,\psi(x, y, z)]$$

where P = pressure at the point (x, y, z)
U = source strength
$\psi(x_0, y_0, z_0)$ = room mode at the source position
$\psi(x, y, z)$ = room mode at point (x, y, z).

Doig [19] applies this theory to mitred bends and defines the ratio of the amplitudes of the ingoing and outgoing wave forms as the coupling factor. Because of the similarity between mitred bends and tee-junctions, Doig is also able to apply his work to the latter situation. One problem Doig finds, however, is that he has to assume equal energy in each ingoing mode. This is obviously an over-simplification in the context of an entire ventilating system, as it ignores the effect of the system prior to the fitting being considered.

Figure 4.3 shows a continuous traverse from one side of a duct to the other, before and after a mitred bend, as obtained by Redmore [21]. The sound source was a discrete-frequency input at the frequency shown. As the duct was 0.3 m square, it can be calculated that the mode of the lowest order should propagate at around 570 Hz. Thus, a plane wave would be expected to propagate at the lower frequencies, 250 and 500 Hz, as shown, but at higher frequencies the influence of the modes can be seen.

Figure 4.4 shows some modal frequencies before and after the bend. As can be seen, the bend alters the relative modal strengths just after the bend compared to just before it, and this can be expected to influence the attenuation along the side branch. The distance over which the modes retain an effect depends on the initial strengths and the decay rates of the individual modes, compared to the plane wave.

Figure 4.3 **Traverse across a duct at frequencies shown for (a) before the mitred bend, (b) after the mitred bend [21]**

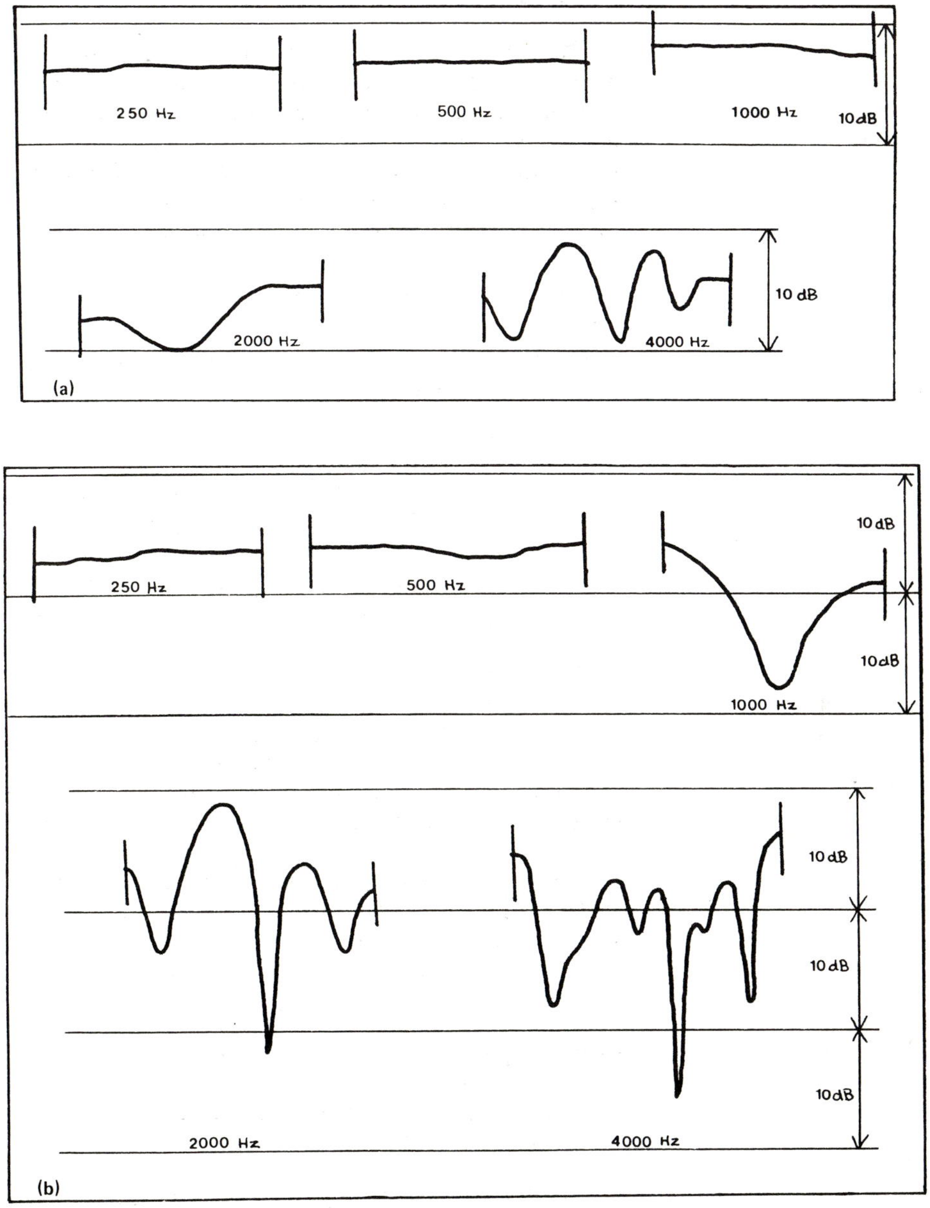

Mode-coupling theory has already been used to explain the existence of cross-modes in branch ducts and has shown that, contrary to existing prediction methods, the important dimension is the one in the plane of the bend and not the minimum. This is because the cross-modes in the plane of the bend (i.e., the horizontal) are excited much more strongly than those in the other plane (i.e., the vertical).

Doak [22] discusses in detail the propagation of sound modes in hard-walled ducts, with respect to various sources, while Osborne [23] considers mode propagation in short curved bends. Hence, it can be seen that mode theory can be applied to various areas of ventilating systems.

One problem of mode theory is that it requires complicated calculations to provide practical solutions and, therefore, tends to be left as a theoretical solution. The increasing power and use of microcomputers, however, means that theories that previously would have been used only to provide limited solutions to those having access to main-frame computers can be incorporated in programs and used by the increasing number of people who have the use of a microcomputer. Redmore [21] is continuing and further develop-

Figure 4.4 Traverse across a duct at approximate modal frequencies shown for (a) before the mitred bend, (b) after the mitred bend [21]

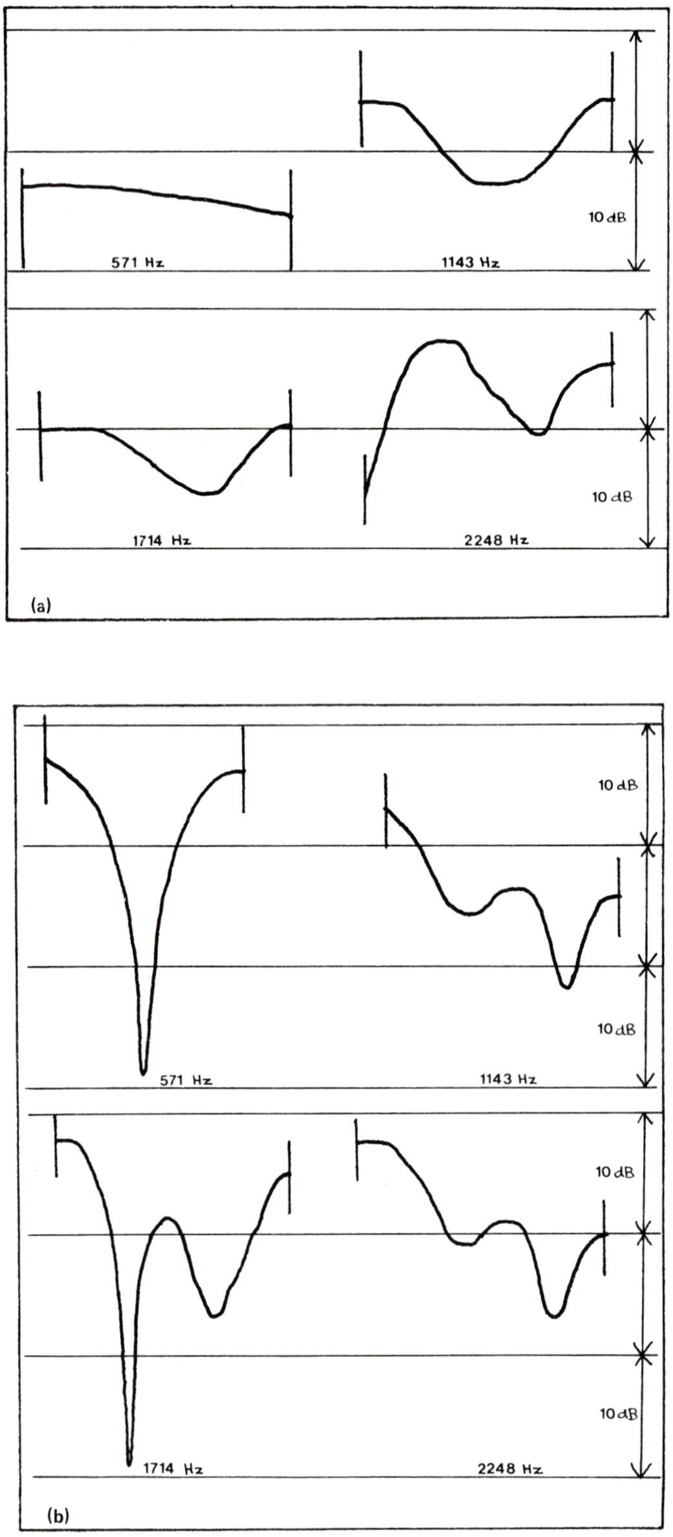

ing the work of Doig, to take into account other areas of duct systems, with the aim of producing a new approach to the prediction of noise in ducted air systems. As the prediction method will have a theoretical background and will not be dependent upon empirical data, it is hoped that this will lead to a better description of the behaviour of sound in duct systems.

The use of computers as time-saving tools has the advantage that calculations involving small amounts of attenuation or regeneration can be included, whereas previously their inclusion may have been left to the discretion of the engineer. Atkins [17] already have computer programs to calculate regenerated noise, and make the point: 'extensive calculations are required in order to arrive at the velocity generated noise power level and these can be very laborious if a lot of them have to be performed manually'.

Croome [2] found the traditional belief that regeneration does not occur below 10 m/sec to be incorrect. Results from tests with variable fan speeds found that sound generation is just as likely to occur at low speed as at high, one effect of increased air velocity being an increase in distance before a bend at which sound generation occurs. Croome's work is, also, in agreement with that of Doig, in finding interaction between elements located close together.

A closer theoretical investigation aimed at a practical solution of regenerated noise in duct sytems is required. This will be included in the work of Redmore [21].

3.3 Cooling towers

Although cooling towers are an external source of noise, they are part of the mechanical services and are, as such, designed and installed for the building to which they are liable to produce the most nuisance. Because of the external location, however, noisy cooling towers can lead to complaints from neighbouring properties.

There are two areas from which sound may originate. They are:

(a) Mechanical noise from the fan, motor or bearings

(b) Circulation noise from the splashing of water, movement of air and rattling of casings.

The overall sound power level can be estimated from:

$$L_W = 11.5 + 10\,\lg P$$

where P = total rated fan power output (kW).

Correct siting of a cooling tower can often determine whether it will be a nuisance or not, so consultation at the planning stage is important. If problems are expected, rather inefficient methods such as using fibrous baffles or oversized cooling towers will lead to noise reduction.

3.4 Compressors

Noise problems due to compressors used in refrigeration and cold storage equipment often arise through bad siting or poor isolation. Centrifugal compressors produce noise at the blade-passage frequency f_b:

$$f_b = b \times n$$

where b = number of blades
n = fan speed (rps)

while pure tones in reciprocating compressors are produced at the piston-stroke frequency f_p.

$$f_p = n \times c$$

where c = number of cylinders
n = revolutions per second.

Croome [24] emphasises the need for good decision-making in the control of noise from compressors, in particular careful selection of equipment; co-ordination between the design team in equipment location and the choice of building materials; and the use of sound control devices.

Acoustic enclosures with one side open reduce the transmission of air-borne sound, while care should be taken to isolate the components from both the structure on which it is sited and the system to which it is connected. Consultation between the designer and the manufacturer at an early stage is important, as noise spectra vary with different models, as do the radiation patterns.

3.5 Boilers

The IHVE Guide states that the sound level (dBA) at 1 metre from a fuel burner is:

$$L = 12.5 \lg Q + 20$$

where Q = thermal output of burner (kW).

Noise originates from the turbulence in the flame zone, and occasionally from instabilities in the fuel or air supplies.

Cummings [25] discusses combustion noise, in particular turbulent noise and noise from inherent system instabilities. Besides describing the noise generating mechanisms, Cummings reviews the current understanding of the problem and summarises preventative and curative measures. These measures can be summarised as:

1. Turbulent noise: Although the requirements for quietness and satisfactory combustion appear to oppose each other, a compromise may be reached by considering the following:
   (a) optimum burner design would lead to velocities as low as reasonably possible
   (b) different burners with the same fuel produce different noise generation
   (c) a series of small flames tends to be quieter than one large flame.
2. System instabilities: Pulsations in the fuel or air supply systems can be removed by changing the equipment, or by fitting attenuators. Other inherent instabilities are harder to eliminate and it may be necessary to use a trial-and-error method for removal.

A discussion of some practical aspects of noise from boilers and burner units is presented by Field [26]. Among methods suggested for noise reduction are the alteration of the burner geometry or the fuel supply technique; the use of acoustic shrouding to reduce the amount of noise radiated into the plant room; and acoustic traps to reduce the noise emitted through the flue. Field also noted the fact that the frequency distribution, in particular peaks, depends on the boiler surface geometry, and can be used to shift the position of resonances, and thus obtain a better sound spectrum in relation to the dB(A) scale.

Although acoustic enclosures can be used to reduce the sound output, they will not cover air-paths such as chimneys. The sound level (dB(A)) in chimneys can be predicted by [24]:

$$L = 9 \lg Q + 45$$

where Q is the boiler rating in kW.

As any noise emitted from a chimney will be into the open air, it may be necessary to use attenuators to reduce the output. It is also worth considering that a chimney allows a path for external noise to enter the building. Duerden [27] refers to experiments carried out near London Airport on helicopters flying over houses that found sound-pressure levels 6 dB(A) greater in a room with an open fireplace than when the fireplace was closed up.

3.6 Motors

All electrically driven motors emit an induction hum at twice the supply frequency (100 Hz in the United Kingdom). Noise can also be produced from magnetically induced vibrations in the region of 500–1000 Hz.

Mechanical parts such as bearings and brushes can be noisy, especially if the machine is badly maintained or becomes subject to imbalance. Also, ventilation of the motor to stop overheating can lead to airborne noise.

3.7 Pumps

Except through poor maintenance or incorrect installation, noise from pumps should not generally be a problem. As with a compressor, a fundamental tone, the blade-passage frequency, f_b, is emitted, together with harmonics. Correct inlet design reduces shock and turbulence in the fluid, while flexible connections should be used on both the inlet and the discharge to minimise flanking transmission.

Noise can also originate from mechanical and electrical parts. Any dirt on the impeller, or corrosion or wear of the pump parts can lead to imbalance, with an increase in noise.

Although vibrations at the pump may not be noticeable there, they might produce resonances in other parts of the system.

From a survey of more than 400 complaints concerning pump noise, Clark and Petrusewicz [28] found that, although measured noise levels were usually very low, the noise spectrum usually contained 2 or 3 discrete-frequency peaks in the range of 300–1000 Hz. If the predominant noise was within this frequency range, people did not become accustomed to the noise, and were only satisfied if the noise emitted from the pump was reduced to a certain level. They also found that most of the complaints referred to noise coming from the radiator surfaces, or other structural parts, and in one instance the noise was radiated from a surface at a distance of 60 m from the pump.

3.8 Pipes

There are a number of causes of pipe vibration. The blade-passage frequency of a pump may excite a section of piping, or insufficient support may allow resonances in other parts of the system to excite the pipe. The resonant frequency of a pipe is a function of its length, the velocity of sound in the fluid and the end conditions. The fundamental and harmonic frequencies of a closed pipe are given by:

$$f = \frac{V}{2L}, \frac{3V}{2L}, \frac{3V}{2L}$$

where L = length of pipe (m)
V = velocity of sound in the fluid (m/sec)
f = frequency.

For a pipe open at one end this becomes:

$$f = \frac{V}{4L}, \frac{3V}{4L}, \frac{5V}{4L}.$$

Changes in pipe diameter, bends and fittings affect the resonant frequencies, while pipe insulation can damp it out.

The effect of resonance on the noise produced in the pipework of water systems is emphasised by Mackensen [29]. He suggests three basic design parameters to reduce the chance of resonance:

1. The pump vibration frequency and the natural frequencies of the circuit should not be the same.
2. The amplitude of the pump vibration should be kept as low as possible.
3. The components of the circuit should have good damping properties and the range of natural frequencies should be limited.

Mackensen also discusses how noise is produced in piped systems, methods of reducing the transmission of sound, and how noisy installations can be checked to locate the source of the noise.

Air that has not been vented off at the filling stage, or that originates from low pressure in parts of a system due to incorrect pump positioning can lead to considerable noise. Gas from corrosion processes can have the same effect.

3.9 Fittings in water systems

Fittings such as valves, reducers and air vents frequently create the conditions within water flow to generate turbulence, vortices and flow interruptions. A sudden interruption, such as a stop-valve closing, results in an extremely sharp pressure rise at that point, due to the incompressibility of the fluid. This pressure change leads to a steep wave front that is reflected back and forth through the system and can produce a loud noise. This is called 'water hammer'.

Restrictions in the flow can lead to localised high velocities and low pressure. Under certain conditions vapour bubbles can be produced which, if the pressure falls below vapour pressure, can erupt with extreme local pressure fluctuations and intense noise. This is known as 'cavitation'.

Ball and Webster [30] found that they were unable to detect cavitation in straight pipes with flow velocities up to 14 m/sec, or in elbows up to 4.5 m/sec. Figure 4.5 shows the pipe vibrations obtained by them for the test specimen shown. Ball and Webster [30] also studied the effect of pipe fittings on noise generated, and obtained the results shown

Figure 4.5 Pipe vibration due to cavitation [30]

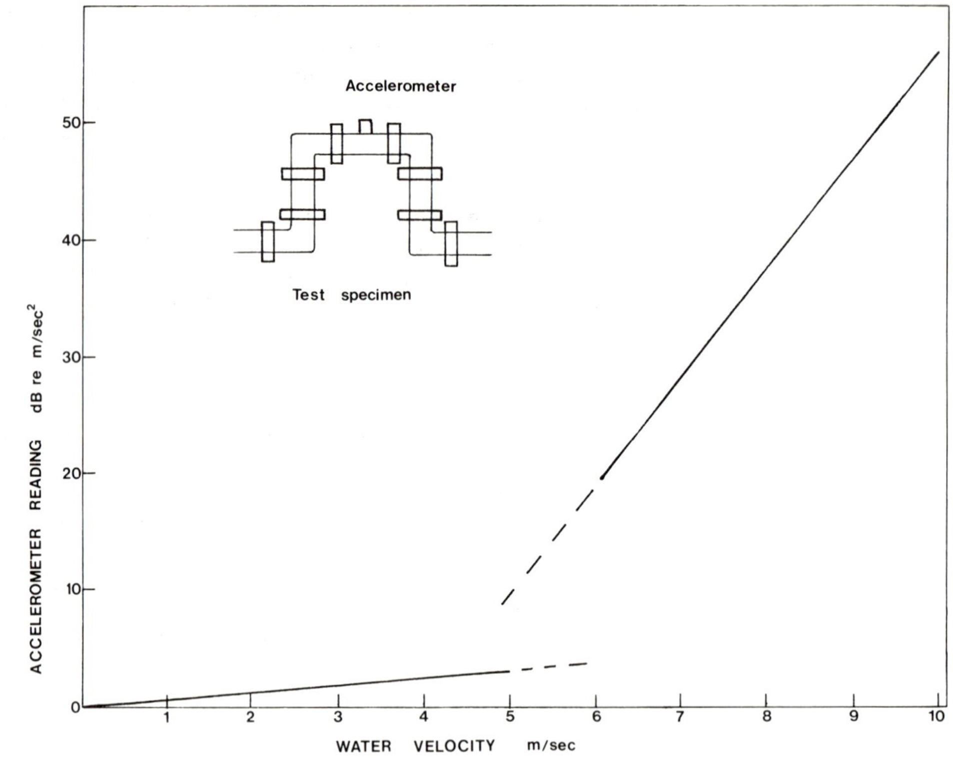

Figure 4.6 Comparison of sound levels obtained with various makes and types of fitting [30]

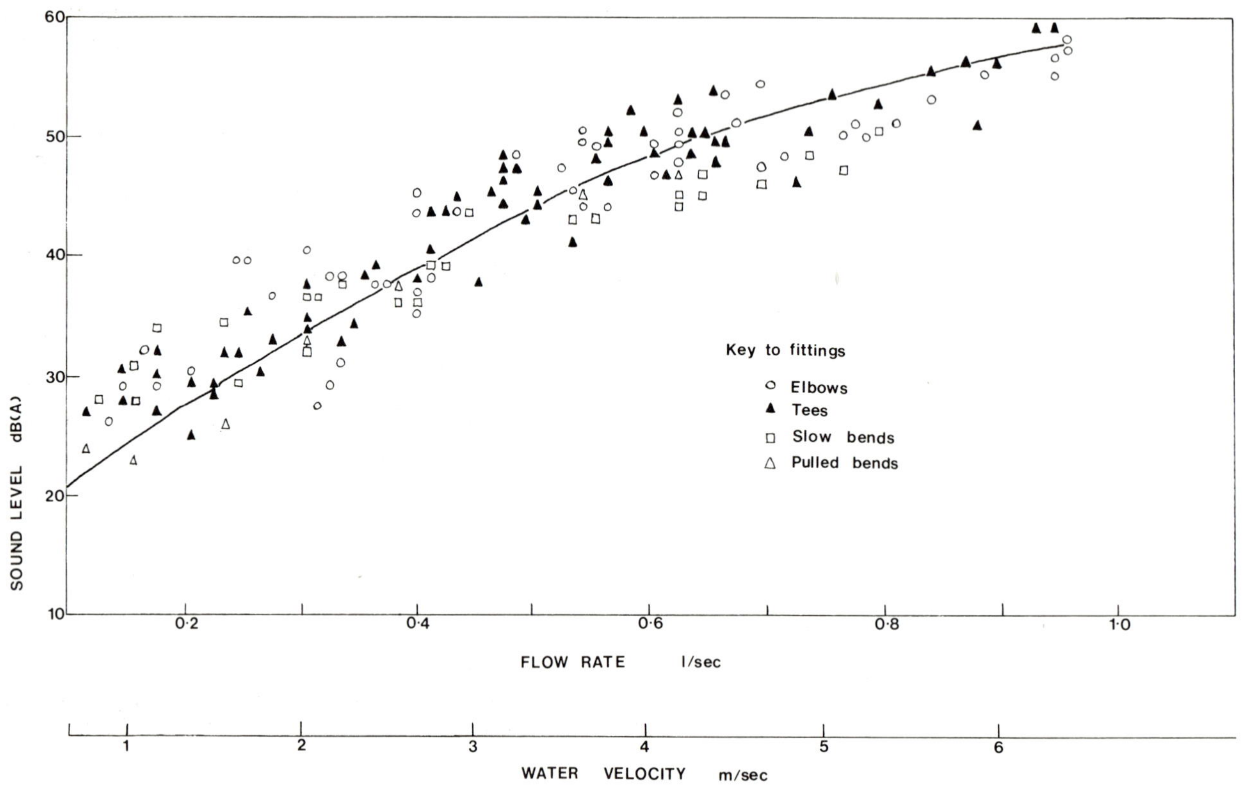

in Figure 4.6. They concluded that the scatter of results they obtained for different fittings at a given rate of water flow was attributable to the differences in the size and shape of the waterways and that although the scatter in the results covered all the tested fittings below 3.5 m/sec, above this flow rate both types of bend were slightly less noisy than elbows or tees. They further concluded that, under normal conditions of temperature and pressure, flow rates of up to 3 m/sec appear to be acceptable when considering cavitation.

The sound level emitted from water systems increases as the flow velocity increases and is inversely related to the pipe diameter. Thus, noise will be affected by the system design, with consideration to the shape and action of fittings, size of pipes and flow velocities. As corrosion within pipes leads to higher sound emission, choice of materials or the use of additives as a preventative measure is important [24].

4. ELECTRICAL SERVICES

As electrical appliances such as motors and fans have been dealt with elsewhere, this section will deal only with lighting and lifts. Lifts may be regarded as a separate communication service, but because most modern lifts are electrically operated and derive their noise from electrical parts they will be dealt with here.

4.1 Lighting

Except for a few occasions when the filament produces noise just before failing, incandescent lamps are silent. On the other hand, there is always some noise from inductive control gear for discharge lighting, including fluorescent tubes. Depending on the ambient level, this noise might not be noticeable although, in some cases, it can lead to an unpleasant hum.

Suspended ceilings or wall panels can act as sounding-boards and amplify any slight noises or vibrations that occur. If the sound level is critical, it may be better to remove the control gear to another area or position it within an acoustic enclosure.

The correct fitting of the gear can play an important role in keeping the sound level low. Attachment of the gear to a solid structure should be considered and emphasis made on firm fitting during installation.

One problem that is not always considered is that light fittings provide a gap in suspended ceilings and allow a transmission path into the void. If the partitions between rooms are only constructed to the height of the suspended ceiling, this sound can now be transmitted through the void to other rooms.

4.2 Lifts

The use of lifts in buildings provides a source of many small, but irritating, noises and an easy path of transmission – the lift-shaft. Care should therefore be taken both in planning the location of a lift and in taking precautions to silence any potential noise sources.

Wherever possible, lift-shafts should not be located next to party walls and should not pass through noise-critical areas. Where this is not possible, the shaft should have its own structure and lining. If the lifting gear is located in a plant-room above the lift-shaft, the room should have a solid concrete floor with holes for the ropes only. All machinery should be isolated from the floor, as well as the lift guide, as this can transmit vibration from the motor via the rope to the building structure. Careful supervision during installation should be considered, as the cost of remedial work such as inserting anti-vibration mountings at a later stage is very great and will result in the lift being out of use during the work.

Control boards can be noisy and should be isolated from the stucture. Metal gates and doors can also be noisy, although doors with good self-closers should be quiet.

5. PEOPLE

The amount of noise generated by people, and by the equipment that they use, depends on the purpose of the building, and the individual rooms. Obviously, a typing pool will be noisier than a library. Typical sound spectra of some building noises can be found in Beranek [12] and Hay [31].

Care at the planning stage is as important to ensure that occupancy noise is not annoying as it is for external or services noise. Location of rooms to protect noise-critical areas and the use of different materials to provide maximum sound insulation or absorption are most economically achieved at the design stage.

Ductwork between rooms provides a path for the transmission of airborne sound and can lead to cross-talk. This can produce annoyance and embarrassment, as private conversations may be unintentionally overheard, or noise may be transmitted from noisy to quiet areas. Good planning, and the use of silencers in the ducts where required, are essential.

Movement of people can lead to noise. Corridor floors are often hard and will transmit impact noise from shoes, trolleys etc. into the building structure. The use of carpets will deaden this noise, although it may be necessary, in some cases, to use floating floor constructions.

Open-plan offices can produce special problems. Movable screens can be used, while telephones with a single bell or a flashing light will reduce disturbance.

6. SUMMARY

This chapter has described potential noise sources in buildings and has included an account of how good planning, construction and maintenance can provide a good acoustic climate. These points can be summarised as follows.

6.1 Planning

Early consideration of the problems that might arise from external noise, plant-rooms, services etc. can influence the layout of the building. Buffer zones should, wherever possible, be included between noisy and quiet areas. The acoustic environment should be determined and the criteria and restraints passed on to the design teams.

6.2 Design

The designers of the various services should be aware of any potential sources of noise in their design. Consultation with manufacturers may suggest alternative pieces of equipment that will provide less noise or make noise control easier. There should be close liaison between the design teams to provide an optimum design and ensure that the designers understand each other's problems. Care should be taken that correct conditions of installation are clearly stated.

6.3 Installation

Good supervision during installation is important. Incorrect fitting of flexible connections, vibration isolators etc. can lead to problems that can be hard to find and expensive to put right. Checks should be made at the commissioning stage that equipment is performing as expected and that the acoustic environment is adequate.

6.4 Maintenance

Many noise problems develop after equipment has been in use some time. The cause of the noise is often poor maintenance – lack of lubrication, dirt clogging air passages, imbalance in motors etc. Often this is taken as a sign that the equipment is getting old and needs replacing. Properly maintained equipment will last longer and provide fewer problems, so designers should ascertain maintenance procedures from manufacturers and pass the information on to the owners of the building.

REFERENCES

1. Doig, R. and Mulholland, K.A. 'A survey of service system noise faults at the University of Liverpool', *Applied Acoustics*, **9**, 265 (1976).
2. Croome, D.J. 'Predicting the sound emission for airconditioning and ventilating systems', *Applied Acoustics*, **9**, 303 (1976).
3. Schneider, W.H. *Air conditioning and acoustics*, ASHRAE, **63**, (56) (1966).
4. Wilson, A.H. *Noise*, HMSO (1963).
5. Road Research Laboratory. *A review of road traffic noise* (1970).
6. Statutory Instruments. *The Motor Vehicles (Construction and Use) Regulations 1973*, HMSO (1973).
7. ISO. Recommendation R 1996 (1971).
8. Parkin, P.H., Purkis, H.J., Stephenson, R.J. and Schlafenberg, B. *London noise survey*, HMSO (1968).

9. Redmore, T.L. 'The effect of traffic noise on noise levels in buildings', Honours Dissertation, Dept. of Building Engineering, University of Liverpool.
10. Kryter, K.D. *The effects of noise on man*, Academic Press (1970).
11. Allen, C.H. 'Noise from airconditioning fans', *Noise Control*, **3**, 28 (1957).
12. Beranek, L.L. *Noise reduction*, McGraw Hill (1960).
13. Finkelstein, W. 'Preliminary computations of fan noise', *Building Services Engineer*, **41**, 268 (1974).
14. Bommes, L. Trans. ASHRAE, No. 2057, 74 (1968).
15. IHVE. *Guide Book B*, Section B12, Curwen Press (now CIBS) (1970).
16. Sharland, I. *Woods' guide to practical noise control*, Woods (1972).
17. Atkins Research and Development. *The control of noise in ventilation systems*, Spon (1977).
18. Gillam, A.J. 'A review of publications concerning service systems noise in buildings', *Applied Acoustics*, **10**, 33 (1977).
19. Doig, R. 'The transmission of sound through ducted air systems', PhD Thesis, Dept. of Building Engineering, University of Liverpool (1977).
20. Morse, P.M. *Vibration and sound*, McGraw Hill (1948).
21. Redmore, T.L. 'Airconditioning noise', SRC Research Contract, University of Aston (1978–1981).
22. Doak, P.E. 'Excitation, transmission and radiation of sound from source distributions in hard-walled ducts of finite length', *Journal of Sound and Vibration*, **31** (1) (1973).
23. Osborne, W.C. 'Higher mode propagation of sound in short curved bends of rectangular cross-section', *Journal of Sound and Vibration*, **45** (39) (1976).
24. Croome, D.J. *Noise, buildings and people*, Pergamon (1977).
25. Cummings, A. 'Combustion noise', *Building Services Engineer*, **41** (9), 123 (1973).
26. Field, A.A. 'Noise in piped heating systems', *Building Services and Environmental Engineer*, **1**, (6), 28 (1979).
27. Duerden, C. *Noise abatement*, Butterworths (1970).
28. Clark, R.J. and Petrusewicz, S.A. 'Noise in domestic central heating water systems', *J. Inst. Htg. Vent. Engrs.*, **40**, A29 (1973).
29. Mackensen, E. 'Noise in domestic pumped hot water heating systems', *Heating and Ventilating Engineer*, **47** (61) (1973).
30. Ball, E.F. and Webster, C.J.D. 'Some measurements of waterflow noise in copper and ABS pipes with various flow velocities', *Building Services Engineer*, **44** (33) (1976).
31. Hay, B. and Kemp, M.F. 'Frequency analysis of airconditioning noise in landscaped offices', *Journal of Sound and Vibration*, **23**, 377 (1972).

5

The role of noise control in building design

Derek J. Croome

> Ordinarily there is not a close connection between the flow of air in a room and its acoustical properties, although it has been frequently suggested that thus the sound may be carried effectively to different parts. On the other hand, while the motion of the air is of minor importance, the distribution of temperature is of more importance, and it is on reliable record that serious acoustical difficulty has arisen from abrupt differences of temperature in an auditorium. Finally, transmission of disturbing noises through the ventilation ducts, perhaps theoretically a side issue, is practically a legitimate and necessary part of the subject.
>
> (W.C. Sabine, *Collected Papers on Acoustics*, Dover, 1964)

1. BACKGROUND SOUND LEVEL IN BUILDINGS

Noise within buildings may originate from the mechanical engineering services (e.g., heating, ventilation and air-conditioning systems, fluid supply and drainage systems); the electrical services (e.g., lighting); the circulation services (e.g., lifts); the communication services (e.g., telephones); process machines (e.g., forging machines, typewriters); people (e.g., impact noise from walking, slamming doors, movement of chairs, cross-talk between one space and another); and external sources (e.g., traffic, aircraft, machinery).

A limited amount of acoustical data is available on these sources, and it is suggested that manufacturers are consulted or acoustical measurements taken on existing similar installations when insufficient data are available to design the acoustical environment. Doig [1] and Croome [2] have carried out surveys which establish that noise can cause serious problems in building design.

The acoustical environment in the occupied space is the resultant of the noise arriving at the space from the engineering services, from adjacent areas by direct or indirect transmission and cross-talk, from the external environment and from noise generation within the space. The building services technologist has a clear responsibility to consider noise problems.

The plant-room may be within or outside the building housing the people. Noise emitted in the plant-room will be:

1. conducted to the occupied space by the fluid distributing the energy or by flanking transmission in the distribution network
2. transmitted through the plant-room structure to adjacent areas inside and/or outside the building.

Further, if the fluid velocities are too high, noise generation may occur in the distribution system and/or at the terminal units.

Noise sources and their control are interrelated. It is the manufacturer's responsibility to produce quiet equipment. The building environmental engineer and the architect need to work together on the noise implications of their building design. Noise control procedure has a systematic pattern to:

1. locate noisy areas away from critical ones whenever possible
2. design good aerodynamic and hydrodynamic distribution systems
3. select plant and equipment carefully, considering the noise implications
4. use building materials and structures suitable for control of noise penetration
5. use sound control equipment, where necessary, only after the above issues have been exhausted.

2. NOISE FROM AIR-CONDITIONING SYSTEMS

Air-conditioning noise is, of course, only part of the background noise spectrum for a space, but it has become increasingly significant as more high-velocity systems are being installed and as people have become more sensitive to internal noise sources in buildings that are well insulated from the external environment. Research by Hay [3] indicates the varying degrees of subjective acceptability in landscaped offices for various differences in sound level between the air-conditioning system and the office with people working in it.

A series of case studies and some laboratory measurements carried out by the author have suggested an approach for estimating the sound emission from air-flow systems which is simpler than the existing method in the CIBS Guide [2]. It also has the advantage of taking into account the interaction between all the elements comprising the system. This research shows that, where there is a strong interaction between any elements of the ventilation systems, any natural ventilation can be cancelled out by generation of sound. The generation of noise at bends varies because interaction occurs. The varying attenuation rates, some are even zero, are also indicative of interaction effects. Doig [4] has confirmed this important aspect of noise patterns occurring in air-flow systems and proposes the use of coupling factor theory to analyse the problem.

In order to ascertain the importance of the problem, a questionnaire was circulated to 49 universities and colleges throughout Great Britain [2]. The questionnaire consisted of ten questions requesting details of any ventilation or air-conditioning services, the acoustic acceptability of the lecture rooms provided with ventilation, steps taken at the design stage to ensure satisfactory noise levels, specific cases worthy of investigation and details of any amplification systems used. Replies from 16 universities were received from architects, and building and planning officers, concerning some 120 lecture rooms. In about a third of these the acoustics were considered unsatisfactory from some point of view by lecturers, building officers or students. Although such rooms are built for the sole purpose of allowing communication between lecturer and student, the inclusion of background noise-level limits in lecture room specifications appears to be rare. In only one instance was a noise level specified: NC 25 was chosen. Nearly half the cases made mention in a vague manner that the noise level should be kept to a minimum.

It is difficult to describe and locate noise sources, but people thought that the fans and motors were responsible for noise in over half the cases, and the air-flow in the ducts in about a third of the cases. Duct-wall resonance, grille noise, structural transmission and traffic noise were contributory factors.

Often the ventilation systems were designed to provide a greater air-change rate when the lecture rooms were fully occupied. This meant that the fan speed increased in proportion to the extra air supplied. Acoustic treatment was not usually recommended at the design stage. Frequently, the system was installed, noise complaints were then received and further money had to be spent to provide acoustic treatment. This usually took the form of sound-absorption material placed in the ducts near the room concerned, or at bends.

Several instances were quoted where the fans were positioned adjacent to the lecture room. Sometimes insufficient thought had been given to plant-room construction, resulting in high noise levels in neighbouring work-rooms.

As a result of the questionnaire survey, it was decided to carry out a site appraisal of the acoustic conditions in 74 of the lecture rooms. The results of the measurements, superimposed on a set of NR curves, are shown in Figure 5.1, and acceptable ratings are given in Table 5.1. In 13 per cent of the cases for single-speed systems the acoustic conditions were intolerable, and in another 21 per cent unsatisfactory.

It is difficult to say which of the rooms classified as moderate would be unsatisfactory, but, according to the standard recommendations, those which exceed NR 30 by more than 2 dB should be classed as unsatisfactory, the 2 dB being the accuracy of the measurements. This means that about 50 per cent of the rooms are acoustically inadequate. The field studies exposed two broad categories of noise source, those originating from the basic nature of the air-flow system and those arising from bad planning or installation.

Figure 5.1 Survey of sound pressure levels in lecture rooms

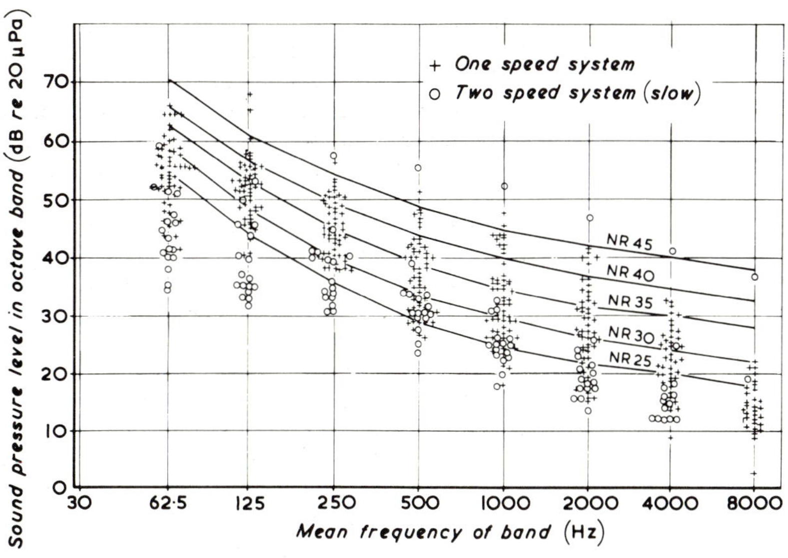

Table 5.1 Acoustic acceptability of lecture rooms (percentage averaged over the frequency range 125–4000 Hz)

Subjective acoustic rating degree of satisfaction	*Single-speed systems*	*Two-speed systems (17 lecture rooms)*	
		Slow speed	*Fast speed*
Satisfactory (< NR 28)	35	81	15
Moderate (NR 28–NR 32)	31	13	17
Unsatisfactory (> NR 32)	21	1	19
Intolerable (> NR 37)	13	5	49

The noise spectra for the plant-room are shown in Figures 5.2 (a), (b), (c) and (d) for two fan speeds. It can be seen that at fast speed fewer peaks are present than for the lower speed. This suggests that, as the speed increases, the broad-band aerodynamic noise in the fan gradually masks the discrete noise from mechanical sources, and even the fundamental fan blade-passage frequency and its harmonics. In this ventilation plant the major mechanical noise sources were the vee-belt and the fan bearings.

Many factors contribute to the noise which can be emitted from vee-belt drive systems. First, the belt travels over pulleys and can vibrate along its passage, due to the formation of standing waves. The fundamental frequencies for this installation are 10 Hz and 3 Hz on low speed and twice these values on high speed: the higher modes of these vibrations will contribute to the low-frequency noise content. Pulleys not running true or belts which are slipping can lead to a variation in angular velocity – the so-called Drehfehler phenomenon. this means that oscillations will occur in the standing wave frequency. A common fault, very prevalent in this site, was due to joints which gave a characteristic belt squeak each time contact was made with the pulley.

Referring to Figures 5.2 (a) and (b), it can be seen that 420–510 Hz discrete frequencies are emitted which correspond to the belt-joint noise. The reason for the wide spread is that three belts are used which will not necessarily all sound at the same instant and, further, each joint will not necessarily cause the same note to be emitted on contact with the small or large pulley. This sound has a high harmonic content which can be noticed at around 900, 1350, 2250 and 2700 Hz.

This plant delivered air to the lecture room of which the ventilation system noise spectrum is shown in Figures 5.2 (e) and (f). It can be seen that at the lower fan speed some discrete frequency components are still present, otherwise the spectrum has a broadband character. The geometrical form of the air distribution duct influences the

Figure 5.2 **(a)–(d) Narrow band sound spectra for centrifugal fans running at 310 and 620 rpm. (e)–(f) Sound spectra in lecture rooms served by the plant-room shown in (a)–(d) [2]**

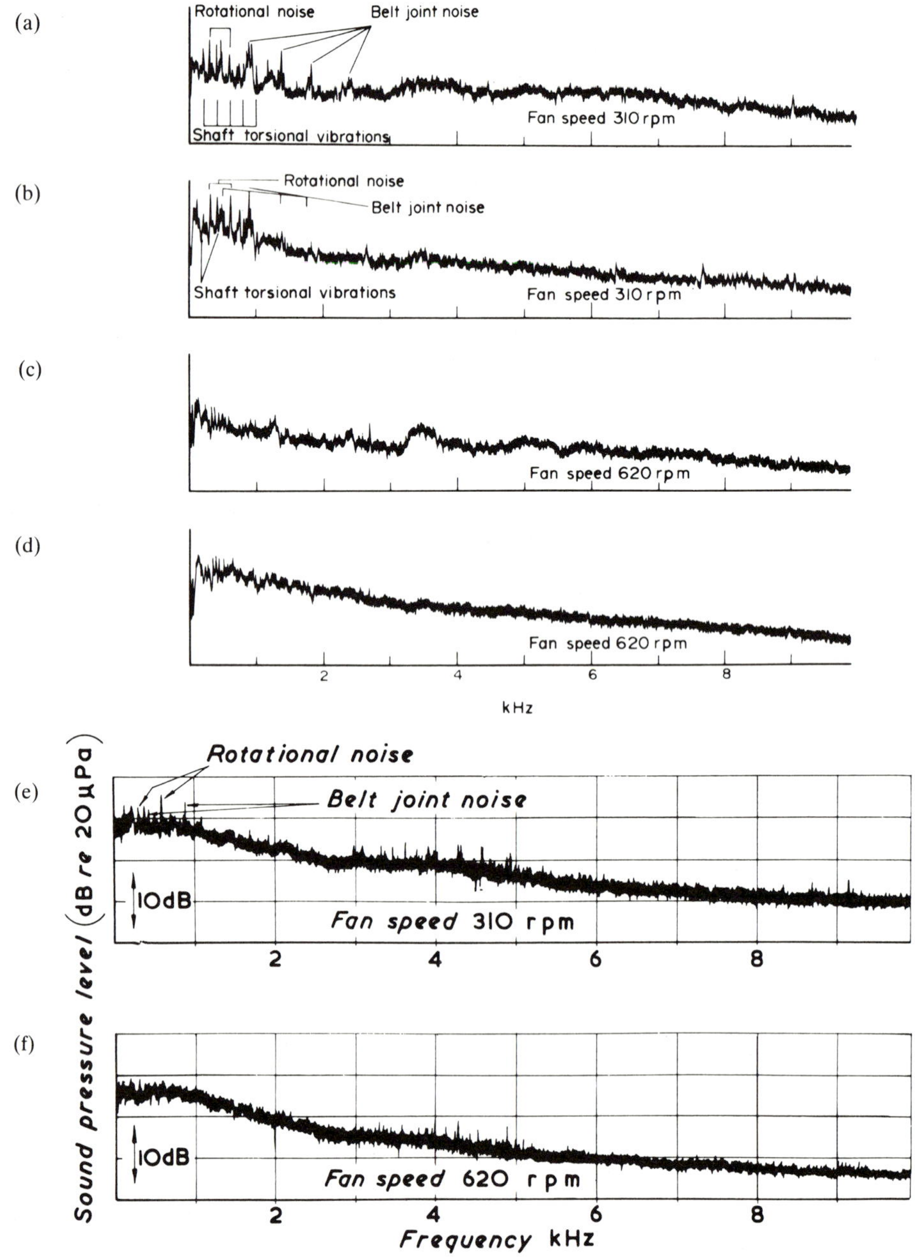

transmission of sound, since bends, expanders, contractions and branch off-takes cause some frequencies to be filtered out.

The investigation was only concerned with air velocities below 5 m/sec (1000 ft/min), where noise generation due to air-flow in the distribution network has, until now, been considered unimportant.

In some cases high noise levels are due to planning which necessitated the positioning of lecture rooms next to a plant-room, or a single-glazed wall adjacent to a main road. Lack of choice concerning available space may force certain environmental factors, in this case noise, to become critical. Commissioning and maintenance procedures must pay attention to noise problems which may arise from faulty installation.

Further research has been carried out on a variable-speed ventilation system serving a 74-seat lecture room in a university. Air is distributed from a unit comprising a 444 mm

Table 5.2 **Performance specification of multi-speed centrifugal fan unit**

	Speed 1	*Speed 2*	*Speed 3*	*Speed 4*
Motor				
Number of poles	4	6	8	12
Hp	1.2	1.3	2.5	5.5
Speed (rpm)	480	720	960	1440
Fan				
Speed (rpm)	760	1145	1530	2290
Volume/flow rate (m^3/sec)	0.73	1.17	1.55	2.33

diameter centrifugal fan with ten backward-curved blades. The drive shaft is mounted on ball-bearings and is driven by a variable-speed motor with four fixed speeds via a double vee-belt and pulleys having a 1.6:1 diameter ratio. Table 5.2 gives the essential performance specification for the multi-speed centrifugal fan unit.

Ductwork is of rectangular cross-section and constructed of 18 gauge galvanised sheet steel; the interior and exterior surfaces are unlined. The heater battery comprises a parallel bank of ten 9.5 mm finned hot-water pipes arranged evenly across the duct in the air stream. The outlet grilles have two rows of aerofoil vanes at right angles: the five of the inner row are fixed, whereas the six of the outer row are hinged and adjustable. The complete system is suspended from a void ceiling by 9.5 mm threaded rods and supported on brackets. The only contact with the void floor is at the grille outlets.

Measurements of the air velocity across the duct cross-section revealed the air-flow profiles shown in Figure 5.3 for the two lower fan speeds only. Air turbulence is evident at the bend between test stations 6 and 7. Later measurements, not recorded on Figure

Figure 5.3 Air-flow profile through ventilation system [2]

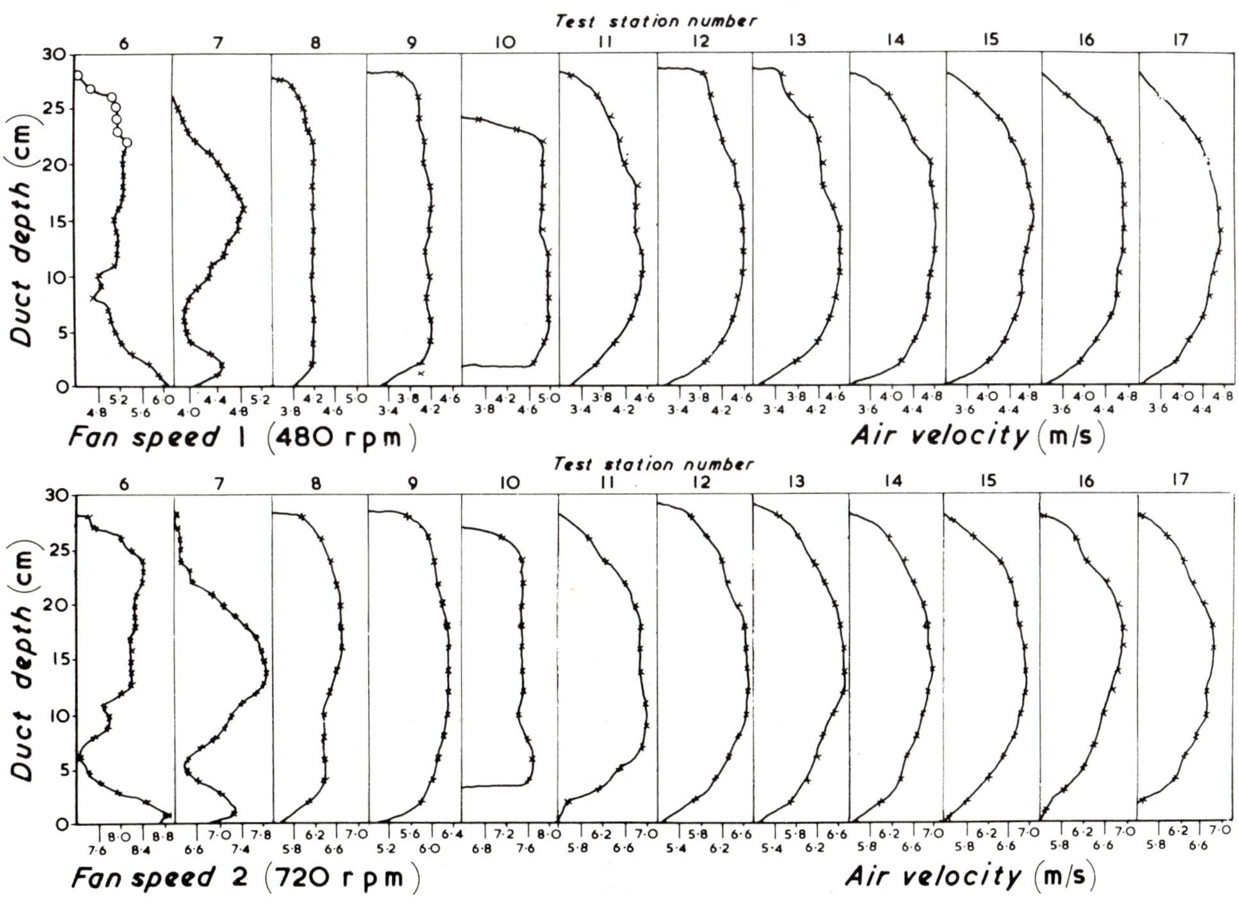

5.3, verified turbulence at test station 18 after the bend following test station 17. The profiles are consistent for the two fan speeds.

Now look at the in-duct sound profiles shown in Figure 5.4. Attenuation of sound occurs at all frequencies but is offset almost entirely by generation of sound which is evident at each bend and at each fan speed, except at the first bend for fan speed 2. The generation happens for all frequencies but is most marked in the 500–1000 Hz range. Sound is generated as the air travels through the attenuator at the three higher fan speeds. One significant conclusion is that sound generation is just as likely to occur in a low-speed ventilation system (average air velocity in this example about 4 m/sec) as in a high-speed system (average air velocity in this example about 13 m/sec). This contradicts the traditional belief that sound generation is insignificant at low air speeds.

It is well established that ventilation system noise has a large low-frequency content. This work reaches a similar conclusion. At the lowest fan speed, sound in the 31.5–25 Hz range has sound levels along the system of 63 to 88 dB (linear), whereas sound in the fre-

Figure 5.4 Sound level profiles measured inside the duct in dB linear [2]

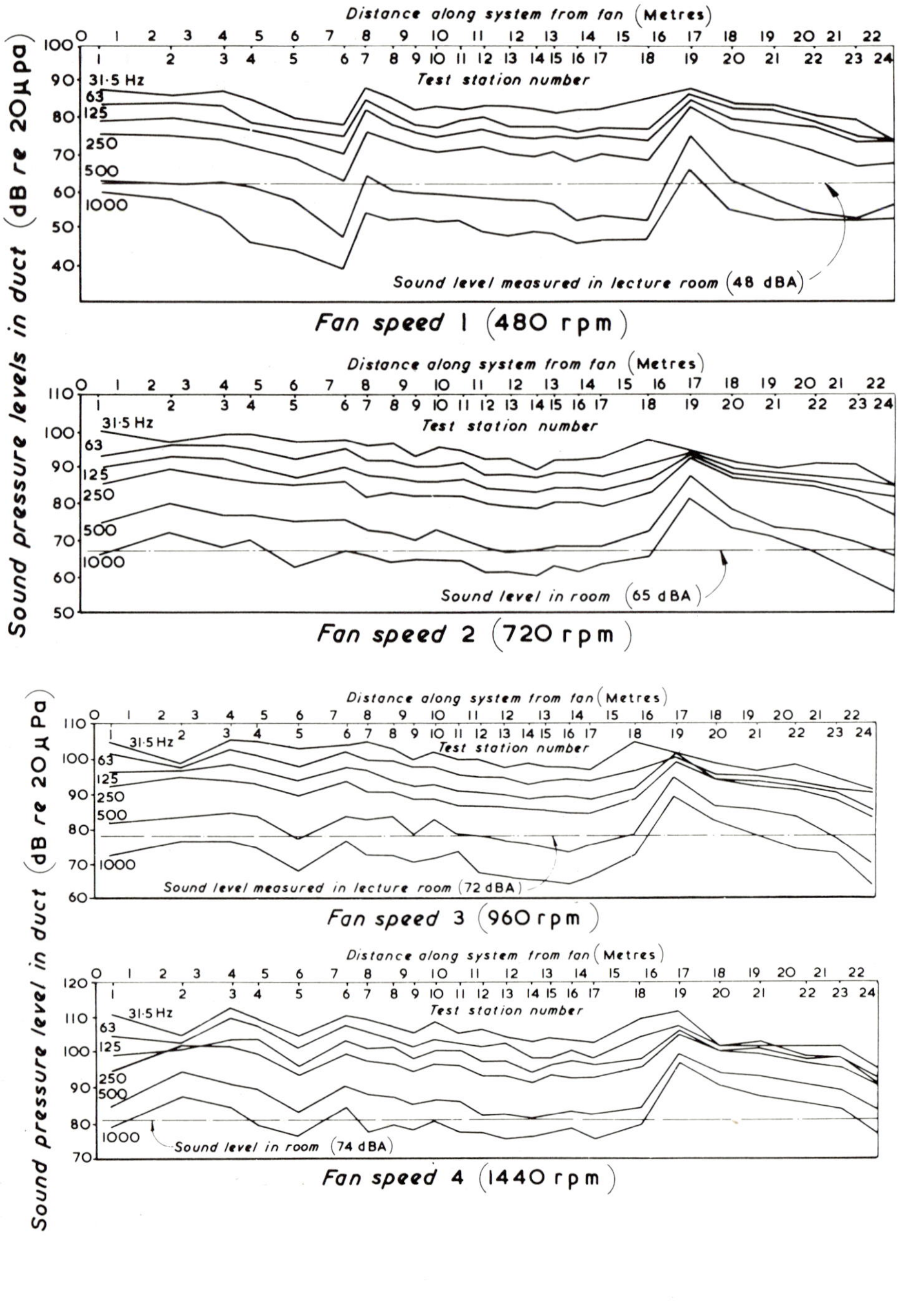

quency range 500–1000 Hz has levels in the range 39–75 dB (linear) (see Figure 5.4). In either case the peak levels are due to sound generation. This pattern of behaviour is similar at the highest fan speed, although the frequencies in the range 500–1000 Hz gradually increase their share of the sound energy as the fan speed increases.

The effect of sound generation begins at some distance before the bend: at low speeds this distance is about 1 m but it extends to 2 m or more as the air velocity increases. The generation at the first bend (test stations 5 and 6) shows some interaction with that which has preceded it at the dampers (test stations 3 to 4), the breeches piece (test stations 2 and 3) and through the attenuator (test stations 1 to 2). Such interaction also happens to some extent between the generation effects at the bends themselves. Duct resonances do not appear to be significant in this system as no fundamental or harmonic frequencies are observed in the spectra.

A very important aspect is that a given rise in sound level due to generation occurs over a much smaller distance than a similar decrease in sound level. This work confirms the point that air-flow systems should be designed, wherever possible, to have the most simple configuration. Good aerodynamic design will assist acoustic design, although a precise relationship between turbulence and sound generation was not established for this system.

The results showed that, where there is a strong interaction between any elements of the ventilation system, any natural attenuation can be cancelled out by sound generation. The generation at the second bend stands out more clearly than that of the first bend (see Figure 5.4), because generation caused by elements preceding the first bend interact with the generation characteristic for the first bend. The varying attenuation rates – some are even zero – are also indicative of interaction effects.

The attenuation rates are variable and very different from those used in practice (see IHVE Guide Book B, 1970). Sound generation magnitudes at bends are of a much higher order – from 4 dB at low frequencies to 18 dB at high frequencies – than attenuation magnitudes in straight metal ducts, which range from 0.5 to 3 dB/m; no attenuation occurs at bends. These results dispute much traditional design data and further research is needed to resolve the conflicts.

A satisfactory design criterion for the lecture room would be an L_{10} level of 36 dB(A) or NR 30. The lowest-speed ventilation produces a sound level of 48 dB(A) in the room. The attenuator makes no effective contribution; in fact, it worsens the situation at the three higher fan speeds.

Figure 5.5 (a) Room sound level related to fan speed [2]

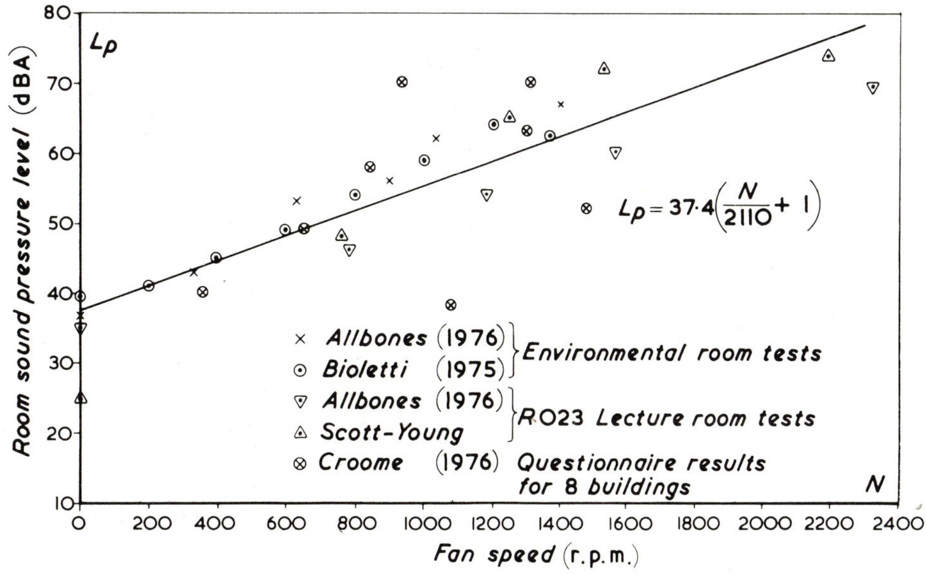

Figure 5.5 (b) Room sound level related to air velocity in main duct [2]

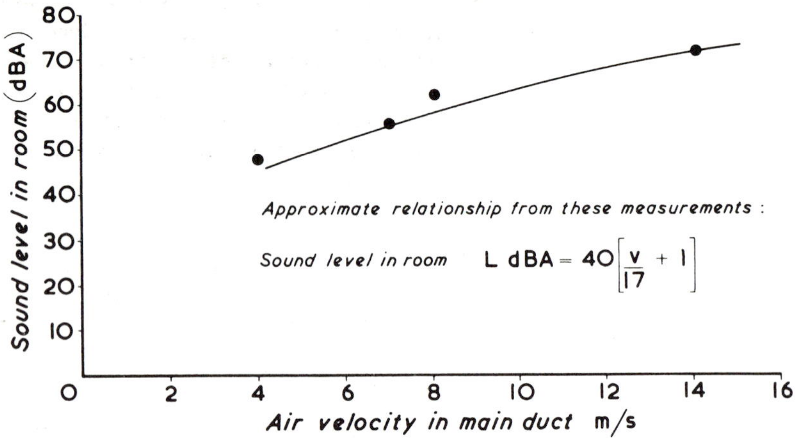

Figure 5.5 (a and b) shows the relationship between room sound-pressure level and fan speed, and between room sound-pressure level and main duct air velocity. It may be possible to predict room sound-pressure level from this type of relationship instead of the more complicated procedures given in the 1970 IHVE Guide mentioned above, but more research is needed in order to establish the influence on this function of system configuration, duct material and various makes of equipment. This could be simply achieved by establishing the relationships derived from such measurements as those given in Figure 5.5(a) for many installations. These preliminary studies show that the sound level in the room, L, can be predicted from the fan speed N (rpm) or the main duct air velocity v (m/sec) by:

$$L = 40\left(\frac{v}{17} + 1\right) \text{ in dB(A)} \tag{1}$$

$$L = 37.4\left(\frac{N}{2110} + 1\right) \text{ in dB(A)} \tag{2}$$

If N and v are fixed by circumstances other than acoustic ones, the order of attenuation required can be estimated. Alternatively, if N can be chosen to match the selected criterion for L, then this value of L can also be chosen to determine v, and hence the size of the main distribution duct can be calculated. Using Darcy's formula (see Chapter 5 of Croome and Roberts [5]), the room sound level may be related to pressure drop per unit length of main duct, Δp, by:

$$L = 40\left[\frac{1}{17}\left(\frac{2gm\,\Delta p}{f}\right)^{1/2} + 1\right] \text{ in dB(A)}$$

for a duct having a mean hydraulic depth m and a friction factor f, where $g = 9.81$ m/sec^2.

A questionnaire has been circulated to 500 firms, requesting them to measure the sound emission from one air-flow system; these results will be entered onto Figure 5.5, and a regression analysis carried out to seek the validity of this simple sound-level prediction method. Some results of measurements made on another system by Bioletti [6] are already included on Figure 5.5.

Further work has established the following conclusions:

(a) Formula 2 is more reliable than Formula 1 and only applies to centrifugal fans.

(b) Natural attenuation and air-flow generation effects are included in the empirical derivation of Formula 2, i.e., the system attenuation and generation which occur between the fan and the measuring point in the room. Additional attenuation, a, may be required and Formula 2 will be modified to:

$$L = 37.4\left(\frac{N}{2110} + 1\right) - a \tag{3}$$

(c) An overall natural attenuation rate of 0.5 dB/m is a conservative estimate which may be used in planning the distance between an air-conditioning plant and occupied spaces but any straight sections in the network which measure less than 7 m in length between fittings such as bends and branches should be ignored when making this estimate. If Lp is the sound level in the plant-room and L in the occupied space, D is the length of air-flow ducts between the plant-room and occupied space, less any straight runs which are less than 7 m, then the sound level in the room is.

$$L = Lp - 0.5D - a \tag{4}$$

Thus, plant selection, choice of fan speed, and distance planning can be decided for a specified value of L. These factors may be combined in such a way that $a = 0$ and no cost for sound-control equipment is necessary, but in practice it is more likely that $a > 0$. Good maintenance is also necessary to maintain the design operating conditions.

(d) The results give reliable predictions (± 5 dB) for centrifugal fan speeds below 900 rpm. Higher fan speeds give less reliable predictions.

(e) German research reported by Eck [7] shows that the loudness level of an axial fan is dependent on speed according to:

$$L = 54\left(\frac{N}{2160} + 1\right) \text{phons} \quad \text{or} \quad 42\left(\frac{N}{1680} + 1\right) \text{dB(A)}$$

(f) Adjustment of louvres in ceiling grilles can have a significant effect on the sound level measured in a room.

3. BASIC PRINCIPLES OF SOUND CONTROL

There is a threefold problem: to protect the building from external noise sources; to ensure that the building designed is not a noise nuisance to people in buildings and spaces nearby; and to control the amount of noise generated within the building.

Increasingly, the building environmental engineer and the architect are being requested to advise on all aspects of the problem. The degree of external protection and internal control required depends on the use of the building. For example, broadcasting studios and concert halls will require a stringent acoustical criterion (i.e., NR < 25). In a factory, however, the permitted noise levels will be much higher (NR 55–58) although in practice often much higher than audiologists would recommend. In designing the indoor noise climate, care is also needed to ensure that people in adjacent spaces outside the building are not aggrieved. The pattern of diagnosis and treatment for these situations is the same, that is, to:

(a) ascertain the acoustical characteristics of the source, propagation pathway and receiver;

(b) provide a mismatch between the source and the receiver, which will be designed from the data obtained in (a) and the acoustical criterion specified for the project. The following sections aim to review briefly the methods available for achieving this mismatch.

Noise control involves employing one or more of the following principles:

(a) *Designing a source to achieve a set acoustical tolerance.* Building environmental engineers have little to say in this matter, which lies in the province of the equipment manufacturers, but they can exercise discretion in selecting fans, compressors and machinery, besides designing efficient aerodynamic and hydrodynamic energy distribution systems.

(b) *Providing a large separation between the source and the receiver.* Making the most of natural attenuation: this involves planning of building sites and also the layout of spaces within the building.

(c) *Enforcing natural attenuation along the source-receiver pathway.* By employing one or a combination of the following:

(i) a high-mass structure to reflect the energy before it reaches the receiver

(ii) an absorption attenuator to dissipate the energy via friction over a broad-band frequency range

(iii) interference attenuators to reduce the sound level at specific frequencies by phase cancellation

(iv) resonance attenuators, which may be tuned to provide attenuation over various frequency band-widths by varying the mass, stiffness and damping characteristics of the attenuator cells

(v) materials or isolators with high isolation efficiency characteristics (i.e., low transmissibility) to minimise the propagation of structure-borne sound.

(d) *Altering the acoustical conditions in the receiving space*. By:

(i) using absorbent materials applied to various surfaces to reduce the reverberant energy level

(ii) using soft floor-finishes to minimise impact noise

(iii) introducing background noise

(iv) isolating the room structure from that of adjacent spaces.

(e) *Restricted work schedules and the use of ear-defenders*. In industrial situations the employment of these will be necessary by workers in acoustical environments equal to or exceeding noise levels of 80–85 dB(A) (refer to Burns [8]).

(f) *Legislation*. This is usually out of the province of the environmental engineer, but refer to the book *The Law of Noise* published by the Noise Abatement Society in 1969 for further information, and also see Kerse [9].

It can be seen that noise control requires a lot of common sense as well as technical expertise.

4. ACOUSTICAL DESIGN BRIEFS AND SPECIFICATIONS FOR BUILDINGS

The brief and specification for a building design must include a comprehensive summary of the acoustical criteria. In order to make sure that the criteria have been selected wisely and can be achieved in practice, an integrated design team working from the design inception to building commissioning is essential. This is obvious when one considers that the required internal noise climate will depend upon the site; the choice of building structure will dictate the relationships between the inside and outside noise climate, and hence the architect and building environmental engineer, together with the structural engineer, must evolve a joint decision on this matter; the orientation of the building will affect the noise exposure of occupied spaces; the internal planning of a building must also take into account the noise-critical areas and place them as far away as possible, or shield them, from external and internal noise sources; various degrees of privacy within the building will necessitate selecting internal wall partitions, ceilings and floors with this acoustical requirement borne in mind; design of services systems has to meet noise level requirements; the choice of surface finishes will affect the sound distribution throughout the space; and in the special case of a concert hall, a theatre or an opera house, the shape of the space must be taken into account when designing the sound distribution. Thus, planning, layout, materials, surface finishes, shape, choice and design of systems are keywords in acoustic design. The problem is not made easier by other factors, such as:

(a) *Cost constraints.*

(b) *The quality of building materials and workmanship of the construction.*

(c) *The interaction of acoustical requirements with other needs*. For instance, the choice of building structure also depends upon the thermal insulation, the solar protection, the daylighting, the natural ventilation requirements, the fire safety, circulation requirements, and the structural requirements; the choice of surface finishes also depends upon aesthetics, cleaning costs and durability; and the choice of the space shape depends upon the social environment and also governs the quality of the air movement distribution, besides acoustic considerations.

(d) *The change of external noise climate with time*. This usually increases from year to year, besides following diurnal patterns.

These problems cannot be overcome without good communication between the client, the architect, the building environmental engineer, the structural engineer and, at a later stage, the builder. Table 5.3 shows a list of the factors which must be accounted for in a building brief and specification in which the noise environment is a priority factor.

Table 5.3 Model acoustical specification for a building space

Building orientation			
Range of sound levels throughout building			
Internal location			
Plant-room location			
Space dimensions			
Volume	m^3		
Roof surface area	m^2		
Wall surface areas	m^2		
Floor surface area	m^2		
Sound levels surrounding space			
External: present		L_{10}, L_{90} dB(A)	
future (20 years ahead)		L_{10}, L_{90} dB(A)	
Adjacent internal areas		L_{10}, L_{90} dB(A)	
Nature of sound sources			
External: traffic		L_{np},*	L_{10}, L_{90} dB(A)
aircraft		L_{np}, NNI, PNdB	L_{10}, L_{90} dB(A)
industrial		L_{np}, CNL	L_{10}, L_{90} dB(A)
people		L_{np},	L_{10}, L_{90} dB(A)
construction		L_{np},	L_{10}, L_{90} dB(A)
future levels		L_{np}, PNdB,	L_{10}, L_{90} dB(A)
Internal: conversations:	with telephone		L_{10} dB(A)
	without telephone		L_{10} dB(A)
machines			L_{10} dB(A)
footsteps			L_{10} dB(A)
plant-rooms			L_{10} dB(A)
General information from client			
Level of privacy required			
Work activity in space			
Work activity in surrounding spaces (including corridor, stairways, doorways, toilets, lifts)			
Machines in spaces			
Number of people in space			
Occupancy hours			
Any future changes envisaged			
Degree of contact needed between employees:	visual		
	aural		
Preferred surface finishes			
Importance of any other aural features			
Selected criteria			
Traffic noise index (external)		L_{10}	dB(A) (18h)
Noise and number index (external)		NNI	dB
Noise pollution level (external)		L_{np}	dB
Background sound level (internal)		L_{90}	dB(A)
Peak sound level (internal)		L_{10}	dB(A)
Sound-reduction index of external structure		R	dB
Sound-reduction index of internal structures		R	dB
Articulation index		0.1 to	0.8
		(privacy)	(good communication)
Design noise rating for services (i.e., assume background sound level is about 6 dB below activity noise level)		$NR \simeq L_{10}$ dB(A) − 12	
Sound-absorption coefficients of each surface		metric Sabines	
Reverberation time at:	1000 Hz	T_{1000} seconds	
	500 Hz	T_{500}	
	250 Hz	T_{250}	
Time delay between direct and first reflected sound		t ms	
Vibration criteria		mm deflection or m/sec^2 acceleration units	
Comments on specific problems			
Services plant-room			
Services distribution system (e.g., damper noise, service voids)			
Terminal noise			
Cross-talk			
Other noise-source problems			

* See Table 5.4 for definition of L_{np}.

5. SELECTION OF BUILDING STRUCTURE TO SATISFY ENVIRONMENTAL CRITERIA

A building structure is not selected for its acoustic specification alone, although in cities and near airports, for example, this is a matter of primary concern. Factors that have to be taken into account when selecting materials for the external building structure are:

Solar protection
Thermal insulation
Condensation risk
Sound insulation
Daylighting requirements
Ventilation rate
Fire protection
Structural considerations
Cleaning, durability and weather protection
Capital and operating costs
Aesthetics

The relative importance of these factors will be different for each building design, depending on the location of the building, its use, and the money available to build and operate it. This section presents some design data to assist in arranging and meeting the chosen environmental criteria.

68 dB(A) on the L_{10} (18 hour) index has been recommended as the upper limit to which residential development should be subjected. This corresponds to about 58–63 dB(A) indoors with windows open. The L_{10} indoors level should not exceed 35 dB(A) in bedrooms at night. The number of complaints increases rapidly when the L_{10} sound level in a general office reaches 60 dB(A); the Wilson Committee set an upper limit of $L_{10} = 55$ dB(A) in spaces where speech communication takes place. For concentrated work in private offices, classrooms and lecture rooms, an upper limit of 45 dB(A) is desirable.

Table 5.4 General internal noise criteria [10]

	Minimum steady (dB(A))	*Optimum steady dB(A))*	*Maximum desirable adverse noise (L_{NP} dB(A)) (a)*		
			Bedroom		*Living room*
Houses					
Rural	25	30	40		50
Suburban	25	35	45		55
Urban	25	40	50		60
Offices					
General	35	45		55	
Private	30	N (b)		55	
Hospitals					
Wards	35	45		55	
Schools					
Lecture rooms	25	30 or N (b)		50 (c)	
Colleges, universities					
Mixed-activity rooms	35	45		55	
Factories					
General areas where communication is important				65	

(a) $L_{NP}\ dB(A) = L_{50} + (L_{10} - L_{90}) + \frac{(L_{10} - L_{90})^2}{60}$ where L_{10}, L_{50} and L_{90} are all expressed in dB(A).

(b) N = 70 - attenuation of partition

(c) Depends on speech interference and size.

Table 5.5 External noise criteria in housing areas

	Urban	*Suburban*
No advantage below these levels	50 (L_{NP} dB(A))	45 (L_{NP} dB(A))
Generally acceptable levels	70	65

Waller [10] has suggested the noise criteria listed in Tables 5.4 and 5.5, which are expressed in terms of the noise pollution level as well as dB(A).

The Noise Advisory Council, concerned about the multiplicity of emission measures used in the United Kingdom for planning and regulatory purposes, has recommended that these should all be replaced by a single measure of environmental noise called the 'Equivalent Continuous Sound Level' (L_{eq}) (for details see *A Guide to Measurement and Prediction of the Equivalent Continuous Sound Level, L_{eq}*, The Noise Advisory Council, HMSO, 1978).

For traffic noise, assuming a Gaussian distribution of levels:

$$L_{eq} = \frac{(L_{10} + L_{90})}{2} + \frac{(L_{10} - L_{90})^2}{57}$$

Aircraft noise in the UK is rated using the Noise and Number Index (NNI) defined by Redmore in Chapter 4, and this may be related to L_{eq} by:

$$L_{eq}\ (12\ \text{hour}) = 0.88\ \text{NNI} + 26.5.$$

There is a corresponding L_{eq} value for the Corrected Noise Level used for assessing noise from industrial premises, details of which are given in the HMSO publication quoted above.

It can be deduced from the above equations that:

$$L_{eq} \triangleq L_{NP}\ \text{dB(A)} - L_{50} - \left(\frac{L_{10} + L_{90}}{2}\right)$$

Table 5.6 shows the sound insulation required for the structure for a given external noise climate and a specified internal noise level. Buildings such as houses, hotels and factories employing night-shift workers require analysis of L_{10} daytime (6 am until midnight) and L_{10} night (midnight until 6 am) conditions; the latter may be the determining factor in some situations.

Table 5.6

	Sound insulation of structure (I dB)		
External noise level	*Internal noise level (L_{10} dB(A))*		
(L_{10} dB(A))	*L_{10} = 35 dB(A)*	*45 dB(A)*	*55 dB(A)*
60	I = 25 dB	15 dB	5 dB
70	35	25	15
80	45	35	25

Sound insulation is defined as:

$$I = R - 10\,\lg\left(\frac{S}{A}\right)$$

for a structure having sound reduction index R, area S and absorption in the receiving space A.

The sound-reduction index does not give the effective sound insulation, because the sound absorption of the space behind the structure plays a part in the sound-energy transmission process. Other limitations to the theoretical estimation of sound insulation are flanking transmission and the quality of construction.

Cracks permit sound to leak through the structure, but at the same time they allow infiltration of fresh air and contribute towards or fulfil, the ventilation rate of an occupied space. (Note: ventilation rate is the fresh air requirement for a space, whereas the air supply rate is the total air supply rate to a space, and includes recirculated air as well as fresh air.) If the external noise climate dictates that sealed windows should be used, mechanical ventilation is necessary to provide the air needed to dilute the carbon dioxide and remove the odours released into the space. Up to about three air changes per hour of ventilation assists significantly in reducing overheating in summer.

The thermal admittance procedure set out in the IHVE Guide, 1970, is used to calculate the level and variation of indoor temperature for various combinations of building location, building orientation, window size, solar protection, thermal response, ventilation rate and heat gain, but noise considerations are also relevant.

Reflective films applied to glazing are more effective than heat-absorbent glasses in minimising the effect of solar radiation on indoor temperatures. Likewise, mid-pane Venetian blinds are preferable to internal blinds, but the minimum overheating will occur with external sunbreakers. The final choice of the solar control system is not easy. There is a high cost involved in cleaning and maintaining external solar protective devices. Reflective films reduce the daylight factor considerably. Reflecting solar energy away from building surfaces is wasteful, but at present there is no economical way of using all the internal building fabric as a solar collector and a thermal store while providing sufficient ventilation and daylight with a minimum of noise transfer.

Wise [11] has proposed a method for assessing the interactive effects of ventilation rate, glazing proportion and type of solar protection on the indoor thermal conditions in lightweight and heavyweight buildings. This method is applicable in situations where the following conditions exist:

Building orientation south-west
Lights not in use
Room heights of 2.4–3.0 m
Ventilation rates occurring in daytime (8.00–18.00 hours BST)
Night-time ventilation rate 0.5 ach
Metabolic rate 60 W/m^2
Local air velocity 0.1 m/sec
Comfort environmental temperature 25°C with a daily range of 8°C for occupants wearing shirt and trousers
Comfort environmental temperature 23°C with a daily range of 4°C for occupants wearing lightweight suit.

(See Building Research Establishment Current Paper CP4/71 for calculation of environmental temperature.)

For other window types, such as tinted and reflective glazing, curtains and roller blinds, it is practicable to interpolate between the limits given using relative values of solar gain factor from Book A of the IHVE Guide 1970.

The design steps for selecting type and amount of glazing compatible with the environmental aspects of building design are as follows:

(a) Select indoor L_{10} sound level
(b) Select window with required external-internal sound insulation
(c) Select peak indoor environmental temperature and daily range
(d) Select daylight factor
(e) Select solar control compatible with conditions in (c) and (d)
(f) Check ventilation rate and see if mechanical ventilation is necessary.

6. ACOUSTIC DESIGN PROBLEMS OF INTERNAL SPACES

Each building presents its own unique problems which can only be discovered as the design is formulated. But there are many problems which keep recurring, and it is these

Table 5.7 Identification matrix for common problems

Acoustical problem	*All*	*Dwellings*	*Hotels*	*Offices, large*	*Offices, private*	*Schools*	*Universities*	*Hospitals*	*Courtrooms*	*Theatres*	*Concert halls*	*Recording studios*	*Television studios*	*Factories*	*Others*
Sources of noise															
External	*	*	*	*	*	*	*	*	*	*	*	*	*	*	
Plant-room	*	*	*	*	*	*	*	*	*	*	*	*	*	*	
Adjacent spaces	*	*	*	*	*	*	*	*	*	*	*	*	*	*	
Machines				*									*		
Typewriter				*	*										
Telephones				*	*										
Duplicating machines				*											
Computers				*											
Footsteps		*	*	*	*	*	*	*	*	*	*	*	*		
Door slamming		*	*	*	*	*	*	*	*						
Heating/ventilation/ airconditioning		*	*	*	*	*	*	*	*	*	*	*	*		
Toilets		*	*	*	*	*	*	*	*	*	*	*	*		
People talking		*	*	*	*	*	*	*	*						
Others	*list appropriately*														
Other problems															
Privacy		*	*	*	*	*	*	*	*						
Deafness														*	
Speech communication				*	*	*	*	*	*	*	*	*	*	*	
Speech reinforcement						*	*		*	*				*	
Echoes				*	*	*	*	*	*	*	*	*	*	*	
Delay time										*	*				
Reverberation				*	*	*	*	*	*	*	*	*	*	*	
Sound quality										*	*	*	*		

problems which are summarised here in the identification matrix shown in Table 5.7, many of them having already been discussed in detail. It should be emphasised that the problems in Table 5.7 are the minimum number that will occur in practice. Some problems can be subjected to a more detailed analysis than others.

Background noise criteria for various types of building are given by the American NC curves and the Preferred NC curves, and the European NR standard curves. The NC and NR curves are similar. The PNC curves have values about 1 dB lower than the NC curves in the 125, 250, 500 and 1000 Hz octave bands; in the 63 Hz, and in the octave bands above 1000 Hz, they are about 4 or 5 dB lower than the NC curves. Concerning the modified PNC curves, Beranek [12] states:

> *The 1957 NC curves have been criticised recently by some users in the USA. It has been demonstrated that if a noise spectrum is deliberately generated with octave-band levels equal to those of a particular NC curve, the sound heard is not a pleasant, or neutral sound, but is both hissy and rumbly. This is particularly true of noise from air-conditioning systems, where duct or fan turbulence adds to the rumble of the random noise sound. To achieve a more acceptable noise quality it sometimes has been necessary to lower the levels by about 5 dB in both the very low and the very high frequency bands as compared to those permitted by the 1957 NC curves.*

7. ACOUSTICAL DESIGN PROCEDURE

For the design and planning of a new building the environmental engineer needs to work in close collaboration with the architect and the structural engineer.

7.1 Planning stage

(a) Assess external noise climate

(i) Measure 10 per cent and 90 per cent sound levels, and the frequency spectrum of particular sources

(ii) Ascertain future developments which may affect the nature of level of the noise climate.

(b) Decide on the design criteria required within the building and plan the layout of the internal spaces, attempting to separate the quieter areas from the noisier ones.

(c) Decide on the sound-reduction index for the external structure, and the orientation of the building, in conjunction with the client, the architect and other members of the building design team.

(d) Decide on the shape of any auditoria, or ceilings in the case of landscaped offices, in conjunction with the architect.

(e) Decide on the sound-reduction index required for the internal walls, partitions, doors, ceilings and floors (also finishes required) with the architect; consult with the the structural engineer concerning the effect of the mass of the building elements on the structure.

(f) Decide on the amount of absorption material required, especially in critical areas such as auditoria, recording studios or landscaped offices, in conjunction with the architect.

(g) Assess the acoustical performance of all equipment to be used in the building (to cover all mechanical and electrical services besides office equipment) in terms of octave-band (or narrow-band) spectra installation or operating device.

7.2 Assessment of noise control equipment required

(a) Estimate dynamic loss required on inlet and discharge side of the system to achieve the design acoustical conditions in the building. (Use method systematically laid out in Book B of the IHVE Guide, 1970.)

(b) Detail final proposals for plant-room structure and any floating floor constructions that may be required.

(c) Check to see if either break-out noise from the system to a quiet area, or the converse, will occur.

7.3 Selection of noise control equipment

Having attempted to obtain the spectrum of attenuation required, select equipment in conjunction with a manufacturer, remembering that interference- and reflection-type sound filters can be easily incorporated as part of the air-conditioning network design, to give the degree of airborne and structure-borne sound control required.

8. THE ROLE OF NOISE CONTROL IN BUILDING SERVICES DESIGN

Building services plant constitutes significantly to the sound level in a building. It is the responsibility of the building services technologist to select services equipment and ensure that its layout in a building is such that the noise criteria are satisfied. But noise is only one energy form comprising the environmental context of a building and it has to be considered along with other climatic factors around the building, such as sunlight and solar radiation.

The background sound level has to be assessed by the building environmental engineer, because surrounding traffic and aircraft noise can mix with services intakes and travel down pipes and ducts. In other words, an integrated approach to noise control is needed, so that the architect receives the best advice, enabling him to select building fabrics and layouts that take into account this aspect. There are strong parallels with thermal and lighting design. The responsibility for achieving the level and quality of thermal comfort and lighting for a particular task lies with the building services engineer, whose role is diversifying and must, surely, include the control of noise in buildings.

The author would like to extend his range of case studies so that a comprehensive catalogue of the noise characteristics of different building types with various services systems can be established. This will form part of a building noise-control guide being prepared by the author for building services engineers and architects. Anyone interested in contributing should contact the author at the School of Architecture and Building Engineering, University of Bath, or at Buro Happold, 14 Gay Street, Bath.

REFERENCES

1. Doig, R. and Mulholland, K.A. *Applied Acoustics*, **9**, 265–79 (1976).
2. Croome, D.J. *Noise, buildings and people*, Pergamon Press (1977).
3. Hay, B. *Phys. Bulletin*, February, 87–8 (1973).
4. Doig, R. 'The transmission of noise through ducted air systems', PhD thesis, University of Liverpool (1977).
5. Croome, D.J., and Roberts, B.M. *Airconditioning and ventilation of buildings* (2nd edn), Pergamon Press (1981).
6. Bioletti, S. Environmental Engineering Finalist Research Project Thesis, Loughborough University of Technology (1975).
7. Eck, B. *Fans*, Pergamon Press (1973).
8. Burns, W. *Noise and man*, John Murray (1968).
9. Kerse, C.S. *The law relating to noise* (1975).
10. Waller, R.A. 7th Int. Congress on Acoustics, Budapest, Paper 245, 10 (1971).
11. Wise, A.F.E. BRE Current Paper CP6/73 (1973).
12. Beranek, L.L. *Noise and vibration control*, McGraw Hill (1971).

6 Building noise case studies

Stephen Edwards

1. INTRODUCTION

This chapter sets out to review a number of projects which either have been successful in achieving the design criteria set down in the brief or have shortcomings for whatever reason at commissioning and have required subsequent alteration to produce satisfactory results.

The chapter also examines the choice of design noise levels for various types of space in which different forms of heating and air-conditioning equipment are installed, and analyses some of the ways in which feedback from case studies has helped to improve design techniques of architects and engineers.

2. CENTRE FOR THE ARTS, POOLE

This is a recently commissioned building comprising a 1600 seat auditorium, a 600 seat theatre, a 200 seat cinema, art gallery, various restaurants and bars. The project is a good example where within a relatively modest budget excellent results can be achieved, and most of the design criteria met. Deep consideration was given to the likely problems of noise and despite the bold siting of the main air-conditioning plant, in a space between the theatre and auditorium, successful results have been achieved. The project incorporates many types of acoustic and anti-vibration equipment to minimise noise.

2.1 Choice of acoustic criteria

Although the acoustic design of the concert auditorium and theatre is not discussed is this chapter, the acoustic performance of the air-conditioning system could not be studied in isolation and the basic acoustic requirements for these two spaces affected greatly the design and choice of the air-conditioning system.

Figure 6.1 (a) Plant-room PNC (to meet auditoria PNC), (b) Auditoria PNC

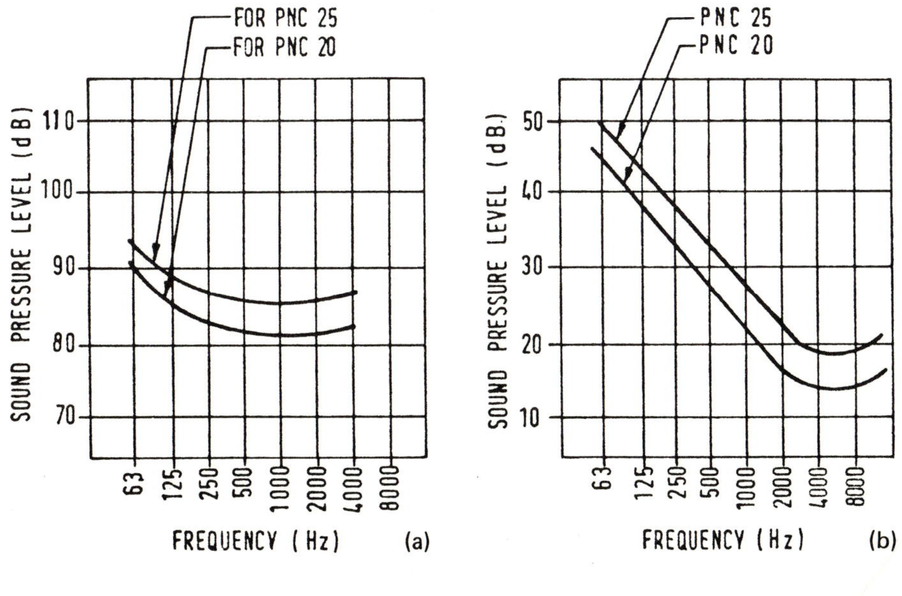

The maximum permissible reverberant sound-pressure levels were chosen for the plant-room and the auditoria. These are shown in Figure 6.1.

2.2 Choice of air-conditioning systems

Air is required to be supplied to these spaces to control vitiation levels associated with high occupancy and to remove heat produced by lighting, other equipment and metabolism. It may be surprising to some to learn how large the quantities of air flow are and how bulky the necessary air-handling equipment becomes. The air-conditioning systems naturally require heating and cooling plant in the form of boilers and water chillers, and ductwork and pipework, pumps and fans are necessary to handle and circulate the various thermofluids. It is the power absorbed by these necessary items of equipment and their thermofluid networks that ultimately produces noise and vibration that may or may not require acoustic treatment. With the criteria set down in 2.1, it is virtually impossible to move the necessary quantities of air and other thermofluids without very careful choice of plant, plant-room design and location, and velocities for the thermofluids. In the Arts Centre the control of noise was a major consideration in the conceptual design and there was excellent multi-professional working from the outset, even during the formulation of the brief. The early joint decision-making on the design produced some interesting and perhaps unusual solutions. These are summarised in 2.3. Figure 6.2 shows a diagrammatic layout of the air-conditioning system, with a typical air-handling unit.

Figure 6.2 Diagrammatic layout of air-conditioning system with typical AHU

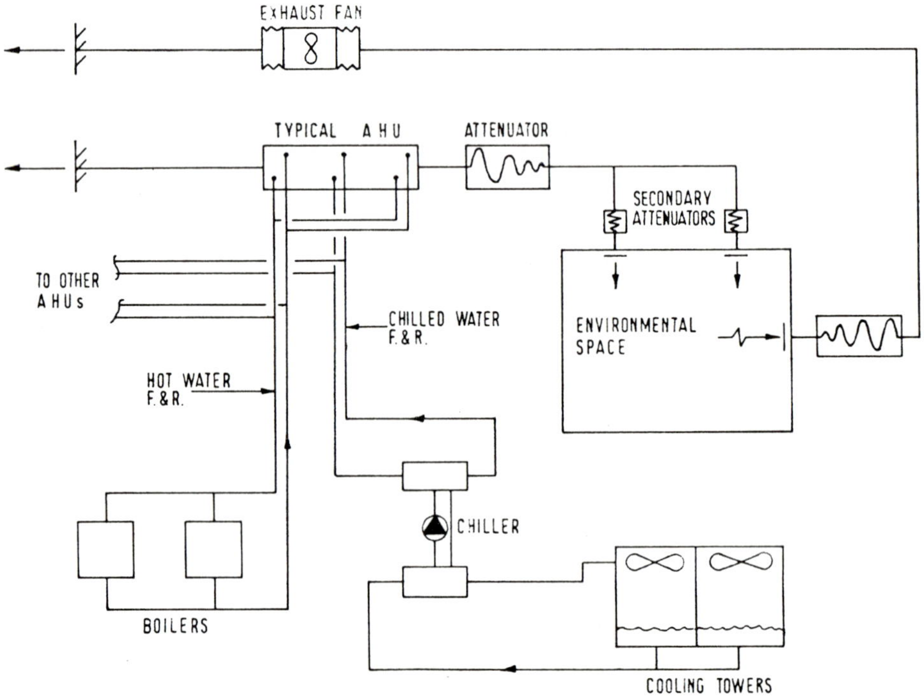

2.3 Special design features

Special considerations that were given to noise and its control resulted in the following design decisions, some of them major.

(i) Water chiller

An absorption machine was selected for its low noise level and anti-vibration, free operation.

Figure 6.3 (a) Frequently used reciprocating or centrifugal compressor chiller, (b) Infrequently used absorption chiller

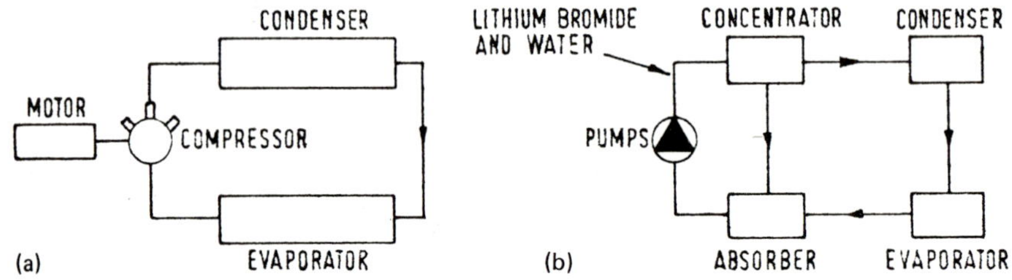

(ii) Two-tier plant-room

The main plant area, housing all the air-handling units (four in all), the water chilling plant, boilers, pumps and supply and extract fans, was located between and directly next to the concert auditorium and the theatre. The simplicity and substantial cost savings in the plant equipment, pipework and ductwork more than compensated for the 'floating' construction of the two-tier plant-room.

Figure 6.4

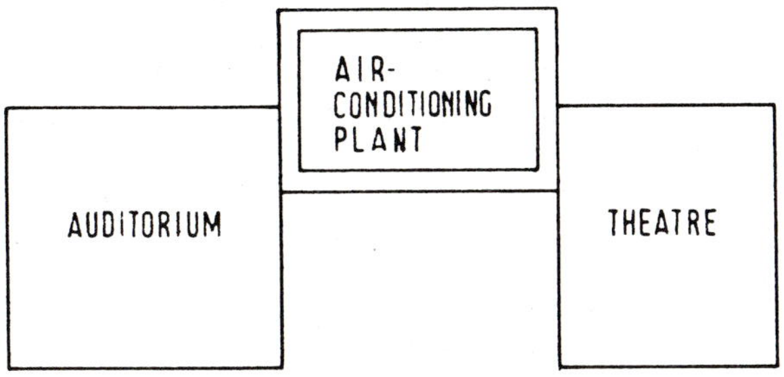

(iii) Anti-vibration equipment

A variety of anti-vibration equipment was selected and considered essential in attempting to meet the rigorous requirements of the PNC levels set for the concert auditorium and theatre.

(a) Inertia bases

Large 'floating' bases have been provided to support large air-handling units and supply fans, boilers and chiller with small floating bases to support, pumps and compressors. Inertia bases, where required, are provided with anti-vibration mountings on outriggers to restrict space below bases to approximately 25 mm.

Figure 6.5

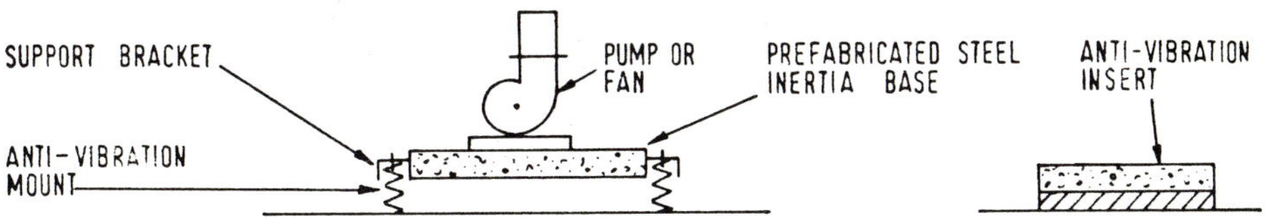

(b) Flexible bellows

Flexible pipework connections to all pumps have been provided in the form of corrugated bellows.

Some engineers consider the use of bellows as marginal, as much of the vibration energy can be transmitted from the pump to the pipework via the fluid. However, in attempting to achieve PNC 20, all marginal savings help.

Figure 6.6

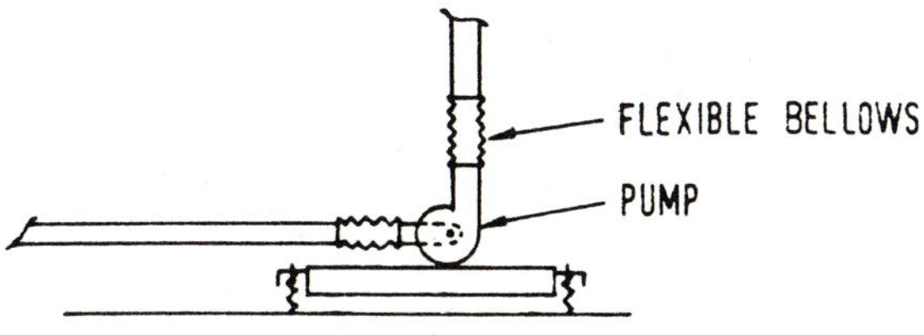

(c) Anti-vibration hangers

Much of the pipework in the plant-room, and all ductwork in the plant-room and voids over the theatre and auditorium, have been supported by rubber in shear or spring anti-vibration hangers. These prevent any residual vibrations in the fluid networks being transmitted to the fabric of the building.

Figure 6.7

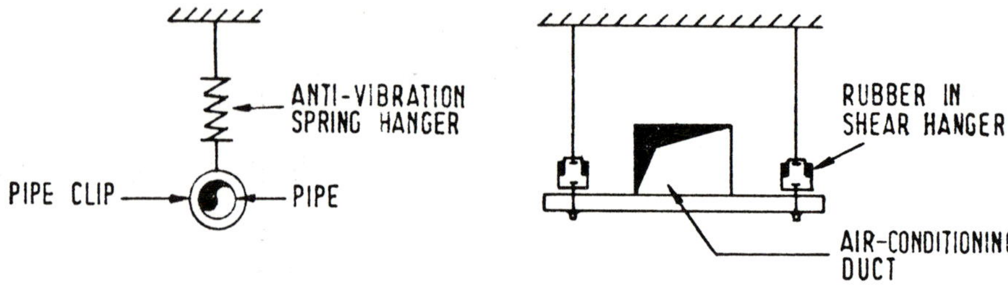

(d) Flexible connections between fans and ductwork

All fans have been provided with heavy duty JPT020 flexible connections, specially selected to minimise noise break-out.

(iv) Acoustic silencing equipment

(a) External acoustic louvres

The fresh-air inlets and exhaust-air outlets of the supply and extract fans have received special treatment to give protection against noise for local residents. The external louvres have been provided with acoustic absorption materials to reduce noise emittance to approximately NR 30 at the site boundary. It was also important to recognise that external noise from the air-conditioning plants could break back into the building. This problem, however, was mainly solved by choosing a fairly high-mass construction for the roof. In one auditorium visited it was possible to hear heavy rain through the roof. Acoustic louvres were also installed around the cooling tower.

Figure 6.8

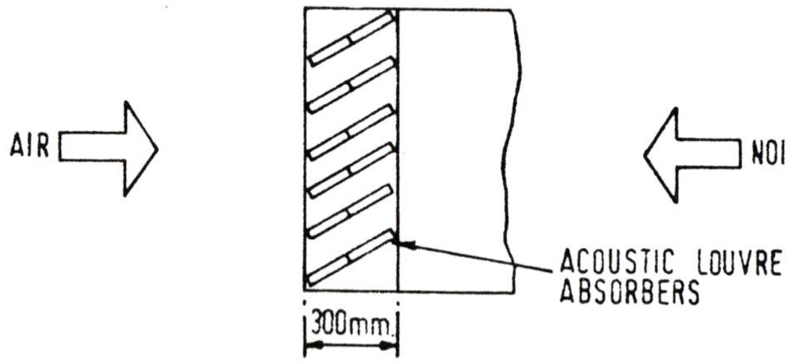

(b) Fan noise-attenuation units

Although care was taken in selecting efficient fans, considerable noise attenuation was necessary and extensive silencing has been added. These silencers are generally provided with acoustic absorber-splitters, as shown diagrammatically in Figure 6.9.

Figure 6.9

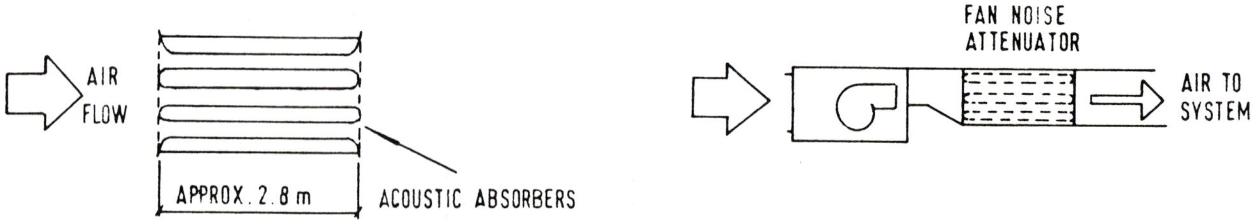

(c) Secondary attenuators

Although most of the fan noise is attenuated by the fan noise-attenuator units, secondary attenuators have been provided at supply diffuser and extract plenum boxes to reduce the possibility of regenerated noise reaching the space. These attenuators, although smaller than the fan noise-attenuators, are similar in construction.

Figure 6.10

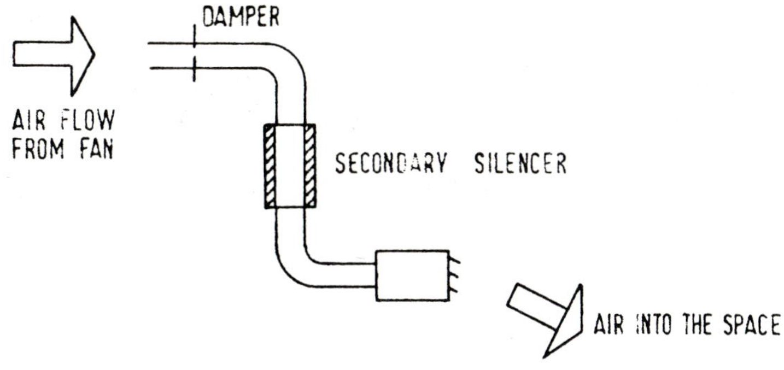

(d) Extract-fan acoustic enclosures

The extract fans chosen for the project were generally of the axial flow pattern. Unlike the supply fans that were contained within the casing of an air-handling unit that was itself acoustically treated, the high-noise-level axial fans not only required attenuators fitted in the air stream, but also required the provision of external acoustic casings to control noise break-out into the plant-room. These casings were of an interesting design. On the inside a perforated metal lining enclosed 75 mm of fibrous mineral-wool absorbing material, the effect of which reduced the overall sound-pressure level in the immediate vicinity of the fan, while the steel outer casing prevented low-frequency noise break-out.

Figure 6.11

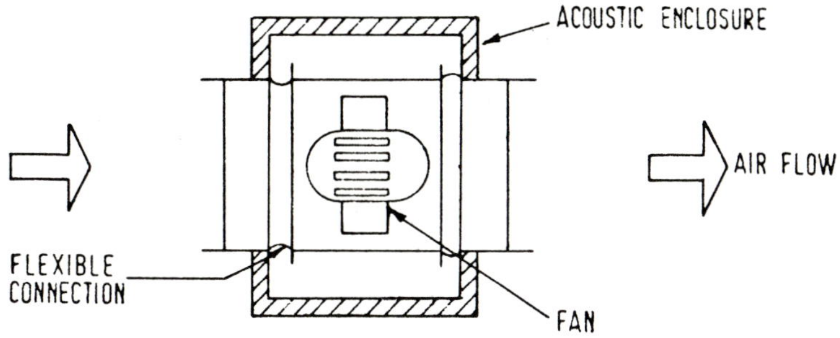

(e) Air-ductwork insulation

Extensive air-ductwork insulation has been applied to all ducts serving the auditoria. The provision of this insulation is primarily for acoustic reasons and consists of a high-density mineral core, approximately 50 mm thick, secured with wire-netting and finished with 10 mm of Keene's hard-setting cement. The insulation reduces noise break-out from the ducts, particularly in the lower frequencies.

Figure 6.12

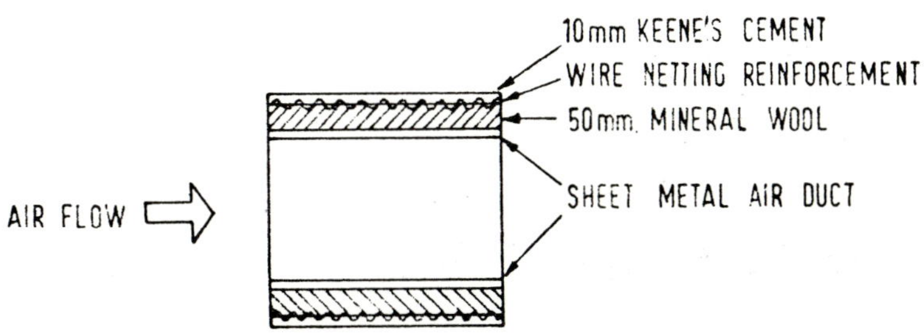

2.4 Measured performance and commissioning adjustments

After commissioning of the various thermofluids systems had been completed, acoustic measurements were taken in the various spaces. First results recorded for the concert auditorium proved very encouraging, and design levels were almost met and were recorded as below PNC 25 but above PNC 20. However, the first results achieved in the theatre were

Figure 6.13 Measured SPLs in plant-room and theatre

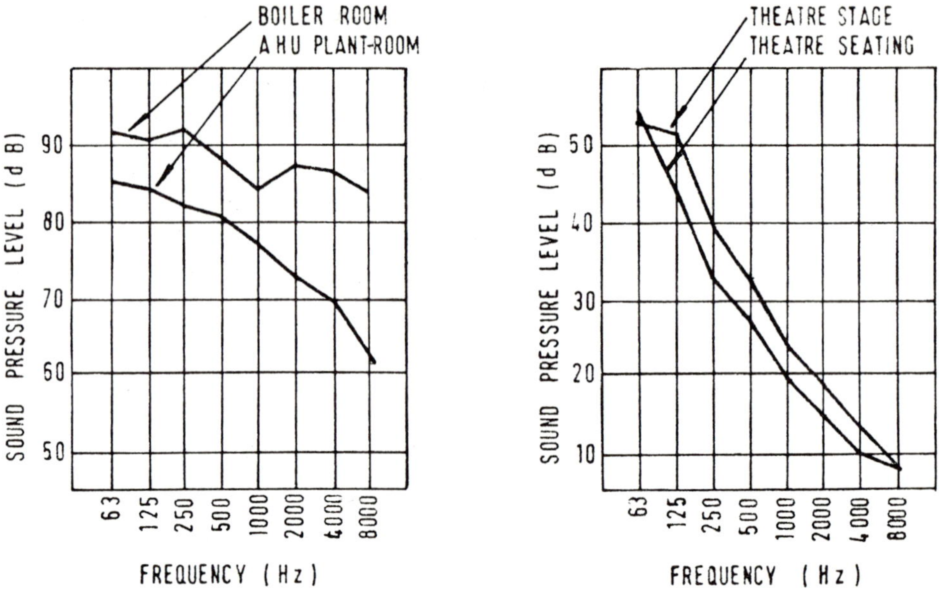

disappointing. Measured sound curves for the theatre and plant-room are shown in Figure 6.13.

Measured SPLs exceed the PNCs for the plant-rooms and the theatre as shown in Table 6.1.

Table 6.1

	Frequency (Hz)							
	63	*125*	*250*	*500*	*1000*	*2000*	*4000*	*8000*
Boiler room	1	8	6	3	–	2	2	–
AHU plant-room	6	6	3	4	–	–	–	–
Theatre seating	7	2	–	–	–	–	–	–
Theatre stage	6	10	2	3	–	–	–	–

Investigations were commenced to find the reasons for the shortfall in predicted noise levels. A number of items were identified, as listed below:

1. *Excessive noise and vibration in the concert hall air-handling plant fan.* This was found to be caused by the resilient mountings for the fan and inertia base system having resonant frequencies that were too close. One of the sets was changed and the problem solved.

2. *Noise and vibration from the theatre air-handling unit.* The cause of the problem was found to be the omission of some resilient spring hangers to pipework and the pre-heater coil pump. These special hangers were added and the problem eliminated.

3. *Vibration of the main chilled-water pumps.* The anti-vibration mountings of the chilled-water pump bases were found to require adjustment. Unfortunately, the pipework installation prevented this adjustment and subsequent pipework alterations were carried out also.

4. *Air-handling unit for general areas.* Noise and vibration emanated from this plant and it was found that one of the anti-vibration mountings had sheared one of its bolts and had 'bottomed out'. Also, masonry rubble was found under the inertia base and was transferring vibration to the structure. The AV mounting was replaced and the rubble blown out by high-pressure compressed air, and the problem rectified.

5. *Gas pipework vibration*. Due to the level of noise and vibration created by the gas-burner booster fan, it was found necessary to support the gas-main close to the burner on anti-vibration mountings. This was carried out and the problem mainly obviated.

6. *Airborne noise in the theatre*. This was found to originate from the main air-supply discharge louvres, although these were acoustically treated. The acoustic louvres were, however, doing the job well for which they were designed, but roof drainage outlets in the vicinity of the louvres had not been properly sealed and an air path existed directly through the roof of the theatre. The gaps were sealed with a dense mastic and the noise inside the theatre from this source disappeared.

7. *A further annoying noise level in the theatre finally remained and was tracked down to the air-supply diffusers*. The noise levels quoted by the manufacturers of the diffusers exceeded design levels and this was apparently caused by the turning action of the blades, necessarily angled to achieve the desired air-flow pattern along the theatre roof. Some adjustment was made to slight advantage.

2.5 Conclusions

Despite the fact that great care was taken in the design stage of the project, from 2.4 it can be seen that many items of plant equipment and building construction required adjustment and correction during commissioning. However, the fundamental design proved successful and the final results achieved, although not quite as good as the design criteria, proved in reality to be very acceptable. It would appear that for projects of this type a degree of over-design is necessary to allow for all the unforeseen small items of noise generation that can result. There is no substitute for careful design and acoustic planning from the outset. In practice it seems that if one requires NR 25, then one must design for lower levels, perhaps NR 20, to allow for all these unexpected circumstances, however unscientific this may appear.

3. SPORTS CENTRE, EAST MIDLANDS

This project has now been in use for two and half years, and most of the noise problems associated with large sports halls and swimming pools have been successfully overcome. Unforeseen problems, however, appeared after the building had been commissioned and emanated from plant noise external to the building affecting nearby residents. The noise was eventually tracked down and additional acoustic equipment installed to deal with the problem. The study is confined to this one problem area.

Figure 6.14 Site plan

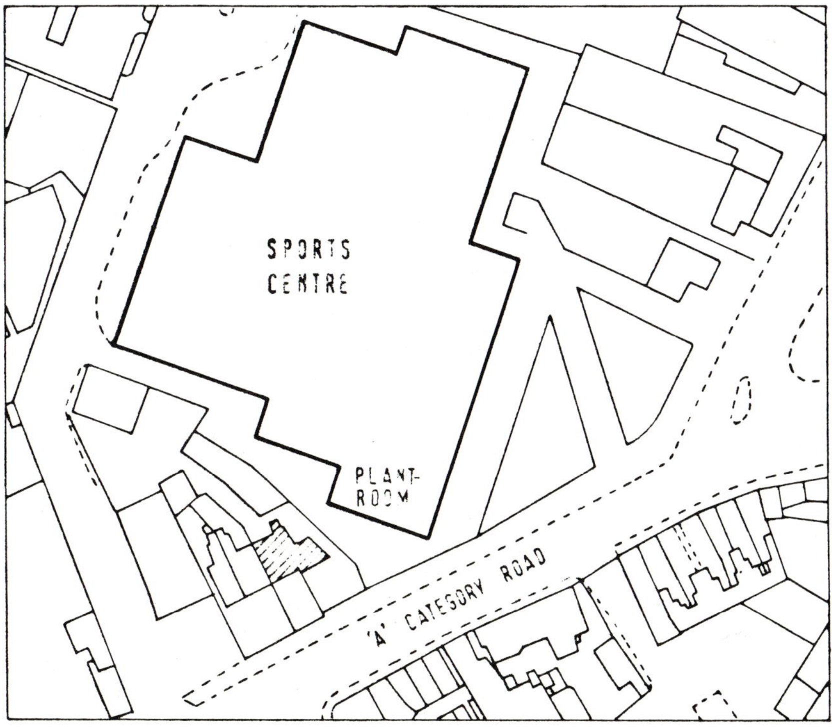

3.1 Site location

The site for the Sports Centre is in the middle of a small town in the East Midlands and is adjacent to a main A road, as shown in Figure 6.14.

The Sports Centre is provided with the main boiler-plant and swimming-pool filtration installation located in the south-west corner of the building and only a few metres from the main A road. On the other side of the road there is a row of houses and a pub, while on the adjacent site three houses are situated. Assessments relating to noise were made during the initial stages of the design and daytime measurements were taken. No extraordinary problem appeared to exist.

3.2 Plant-room layout

The plant-room is constructed on a single level, with access directly from the service road located along one side of the Sports Centre. The entrance into the plant-room for personnel and for equipment replacement or removal faces directly towards the three adjacent houses. The plant-room accommodates two gas-fired boilers, numerous pumps, swimming-pool filters, hot-water storage calorifiers and, partitioned off separately, a chlorine store. A sketch of the layout is shown in Figure 6.15.

Figure 6.15 Plan of plant-room

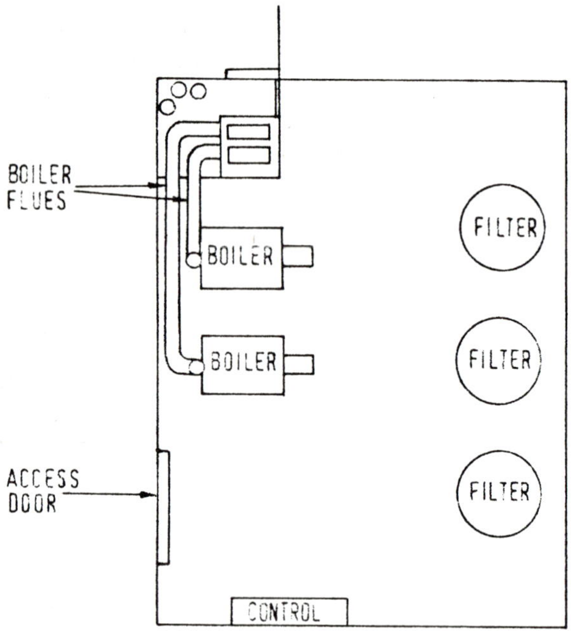

3.3 Commissioning

The total project was commissioned satisfactorily and there were no major unforeseen problems. Certain minor troubles existed but were soon rectified. No apparent noise problems existed and in fact the general noise level throughout the building was very acceptable and in some areas was better than the design criteria levels chosen. As in all buildings accommodating swimming pools, it is necessary to run the plant continuously and switching off at night is not possible.

3.4 The external noise problem

A few weeks after the building had been commissioned and in use, a complaint to the local council was received from the occupant of the house immediately adjacent to the corner of the site in which the plant-room was located. The complaint referred to unacceptable levels of noise and vibration from the plant-room of the Sports Centre. This complaint was unexpected, as measured noise levels from the plant were better than the requirements laid down for 'Residential with Local Traffic, Night Time' either by the Wilson Report or BS 4142:1967. The sound levels measured and required are as follows:

(i) Wilson Report

Recommended levels which should not be exceeded for:

Main residential (night time)	55–44 dB(A)
Residential with local traffic (night time)	53–45 dB(A)
Minor roads (night time)	49–43 dB(A)

(ii) BS 4142:1967

The levels for background noise can either be the actual measured level without the offending plant on or, where this is not applicable or possible, they are as follows:

Suburban, little road traffic	45 dB(A)
Urban residential	50 dB(A)

The plant-noise level measured at the site boundary was 35 dB(A).

The problem in this case study proved to originate from the unusually low background-noise level of the site and the surrounding houses at night, later measured as 23.5 dB(A). This level is normally found in rural areas. So, despite the presence of an A road, with its noise and vibration from heavy vehicles, exceptional low noise levels can occur at night in a town centre location, and quiet plant can then become a noisy intrusion.

3.5 Solution and conclusion

The plant items that were causing the noise of 35 dB(A) at the site boundary were the boilers and their associated gas burners. This noise passed out of the plant-room through the door louvres, which served as air intakes for the burners and were of the necessary free area to satisfy the gas board. Low-frequency vibration was also transmitted from the boilers and steel flue pipes.

It became evident that to eliminate the noise problem, other than by shrouding the burners, which did not overcome the low-frequency noise transmitted by the boilers and their flues, it was necessary to replace the doors with a movable acoustic attenuator. Simple acoustic louvres were not adequate to deal with the problem and a special door incorporating an attenuator was installed, the thickness being almost 3 metres.

Figure 6.16 Section through acoustic door and plant-room

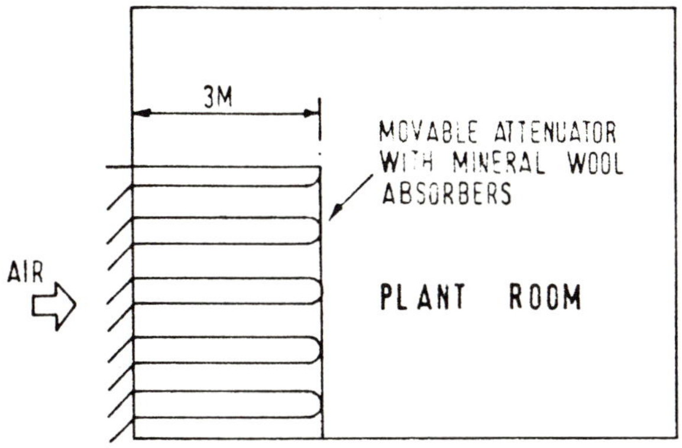

In conclusion, one could argue that this case study is a classic example of where one should always carry out a 24-hour sound survey before embarking on the design of any project. Considering all the buildings that are built, could this suggestion ever be economically justified?

4. WEST MIDLANDS SPORTS CENTRE

This project is very similar in nature to the Sports Centre in the East Midlands but differs in that the latter building was constructed on a site located in a heavily built-up area and alongside an A category road. The Sports Centre in the West Midlands is sited in a rural situation. External noise from air-handling equipment caused annoyance, and residents who were quite some distance from the Sports Centre building were affected.

4.1 Site location

Although the site for the Sports Centre is located close to the centre of a town, the surroundings are very rural. The nearest other buildings are a row of cottages some 180 metres away. Figure 6.17 shows the site arrangement. Close to the Sports Centre is a public car park, which provides general parking as well as catering for the needs of the Sports Centre.

The Sports Centre is heavily biased towards swimming and contains a main pool to national standards, a learners' pool and a separate diving pool. Other facilities include squash courts and sauna suite but no large sports halls have been incorporated.

Figure 6.17 Site plan

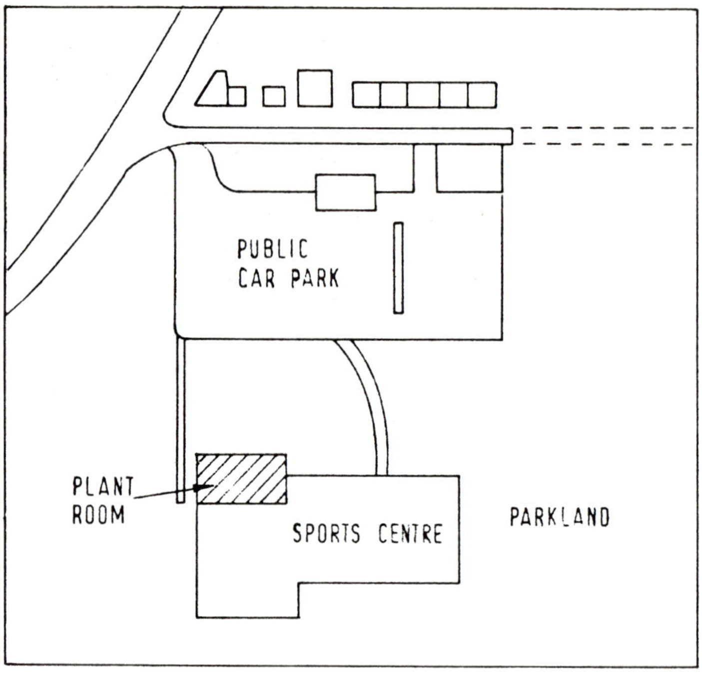

4.2 Plant-room layout

A single plant-room provides space for all the boilers, air-handling units and fans, swimming-pool filtration equipment, pumps and control panels. Natural ventilation louvres are positioned in the walls of the plant-room and face directly north towards the cottages. The plant-room is partly two storeys in height and the higher part houses three main extract fans that discharge horizontally through louvres above the flat roof of the single-storey part of the space. Figure 6.18 shows the arrangement.

Figure 6.18 Plant-room

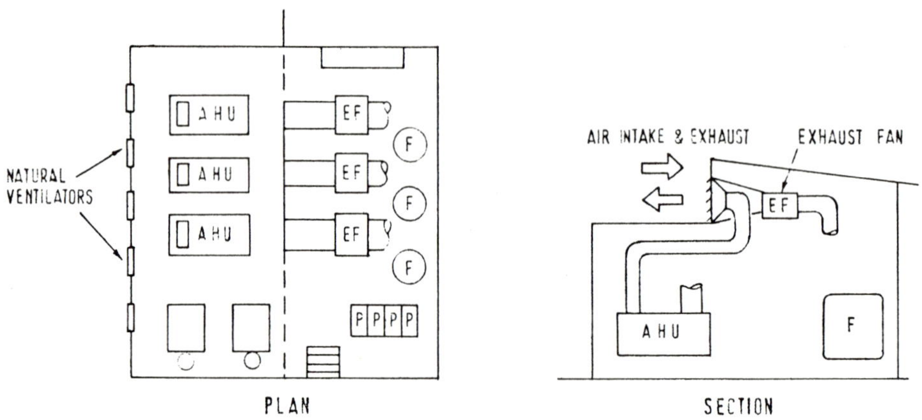

4.3 Commissioning

From the moment the plant was switched on a noise problem was evident and complaints were received from the owners of the cottages. A detailed analysis proved necessary to isolate the various noise sources and identify where the problem lay. Within the plant-room any one item or combination of items could have been the source. Further, the problem could also have been caused by noise passing through the natural ventilators, or through the exhaust or intake louvres in the upper part of the plant-room. A full sound survey was commenced. Similarly to the East Midlands Sports Centre, the plant runs 24 hours and switching off at night is not permitted.

4.4 Analysis and solution

First measurements indicated NR levels of over 68 on the roof immediately adjacent to the intake and exhaust louvres, and almost NR 60 at ground level. The noise level outside the cottages was above NR 40 and was in excess of the allowable NR 30, which should give NR 25 in the bedrooms.

Upon analysing the results, it was found that a highly directional noise was being produced by the exhaust louvres at high level in the plant-room. Switching off the exhaust fans reduced the noise level dramatically and surprisingly the noise level produced through the natural ventilators from all the other plant in operation, including the main supply fans and gas burners, did not exceed NR 30 at the cottages.

The three main exhaust fans, all axial flow running at 2900 rpm, clearly needed attenuating. Attenuators were selected and fortunately, without too much disruption, it was possible to effect accommodation. However, the earliest delivery that could be obtained was some six weeks and this would have had a serious effect on delaying the use of the Centre. Temporary noise control was reviewed and a decision taken to place a baffle in front of the offending louvres. The baffle, constructed of thick plywood and timber framing, proved successful and remained until the permanent attenuators were installed. It became evident also that if the louvres had been originally facing away from the cottages, the noise problem would not have arisen. Figure 6.19 shows the alternative methods of dealing with this highly directional noise problem.

Figure 6.19

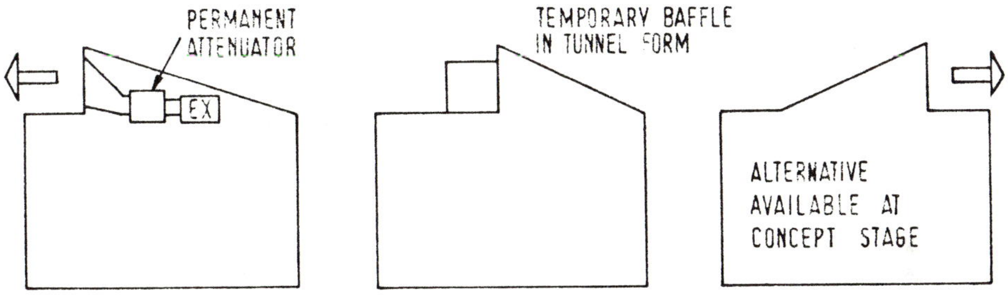

4.5 Conclusion

Firstly, the most important conclusion and perhaps to be linked with the East Midlands Sports Centre is that, where plant is designed to be operated throughout the twenty-four hour period and continuously, a full sound survey is strongly recommended to be made. Secondly, noise from high-speed axial-flow fans can be highly directional and travel greater distances than one might first think. Thirdly, if it is at all possible, louvres should be located facing away from adjoining property. In this case that may not have been too difficult at the concept stage.

5. THE USE OF FAN-COIL UNITS AND INDUCTION UNITS

A number of examples have been chosen for study where various types of noise problem have been experienced. The results of the study appear to indicate that many people are becoming more noise-conscious and rigorous in their assessment of acceptability levels. In one case noise levels from fan-coil units of NR 26 were rejected by a client as too noisy, and the heating system was subsequently changed to a different type. There is evidence also that some manufacturers do not sufficiently research and consider noise from fan-coil and induction units, and are under the misapprehension that building occupants are prepared to accept higher and higher noise levels.

5.1 Thermal requirement and advantages

Fan-coil units and induction units are both capable of satisfying the varying thermal demands of perimeter-type spaces very well, both heating and cooling, and in addition have the facility of easily incorporating the requirement for individual room control. The units are particularly economical in energy conservation and in the air-conditioning application obviate the need for extravagant energy-consuming 'all air' systems.

5.2 Construction of units

Both types of unit employ locally controlled heating or cooling coils, or both. In the case of the fan-coil unit, within the casing small centrifugal fans, filters, the heating and cooling coils and controls are accommodated. Some acoustic absorption and careful fan impeller-bearing design have also been hopefully incorporated. With the induction unit the mode of convecting air over the heating and cooling coils is effected by high-velocity induction nozzles which take the place of the fans of the fan-coil unit. Again some acoustic design has been hopefully applied to the chassis and casing of the unit. In both units the source of noise originates from the fan or the high-velocity nozzles, the air passing through the coil fins and into or out of the air intake and supply louvres. A very few manufacturers have units well designed acoustically, and provide acoustic selection data based upon sound test measurements. Standard units can vary in a well designed unit from NR 28–30 to NR 50–55 in a poorly designed unit.

5.3 Case studies

Case 1: Directors' Headquarters, Yorkshire

This is a high quality project where the use of four-pipe fan-coil units provides heating and cooling. Ventilation is incorporated in a separate low-velocity duct system.

Three-speed fan-coil units were specially chosen for their 'low noise levels' on low speed, as quoted by the manufacturer. They proved to be far too noisy and despite heavy soft furnishings NR levels of between 37 and 40 were recorded and were unacceptable to the occupants. The slow-speed motor position was changed to provide a further speed reduction to an ultra-low setting, only just workable, and the noise levels were accepted with measured values close to NR 30.

Case 2: New House, Jersey

This is a very high quality and ambitious design for an inventor who has a developed awareness of the quality of design. The new building will incorporate a water heat pump with heating water at only 50°C. Fan-coil units for heating appeared to be an ideal way of keeping the dimensions of heating units within reasonable bounds. However, the client was uncertain about tolerance noise levels and agreed to allow a mock-up to be built in his present house in an effort for the consultants to demonstrate the quietness of the quietest units available, if installed under the floor in an acoustically designed box. After installation noise measurements were taken and at the design speed of only 525 rpm the unit generated a noise level in the space approaching NR 25. The background level with the unit off was almost unmeasurable at below NR 20. The client was not prepared to accept NR 25 and the new design incorporates natural convectors only, despite their fairly large dimensions.

Case 3: Office block, West Country

This is an example of an induction-unit installation, with a measured noise level of NR 33 that meets with complete client satisfaction. The scheme incorporates a two-pipe induction system with very carefully selected induction units. The casings around the chassis of the units are provided with sound-absorbing mineral wool and low high-velocity plenum pressures were chosen. Some would say that this example may not make the best economic use of 'tightly sized' induction units but the reliability and absence from any client complaints has proved for the consultants to be significant.

7 Sound insulation: background and recent developments

Neil Spring

1. INTRODUCTION

This chapter gives a brief review of the underlying principles governing the propagation of airborne and structure-borne sound in buildings. Some of the latest developments which could affect the design of buildings are described.

2. DISTINCTION BETWEEN AIRBORNE AND STRUCTURE-BORNE SOUND

Consider the situation shown in Figure 7.1, which is a diagram of a section through adjacent rooms.

Figure 7.1 Sound transmission through adjacent rooms

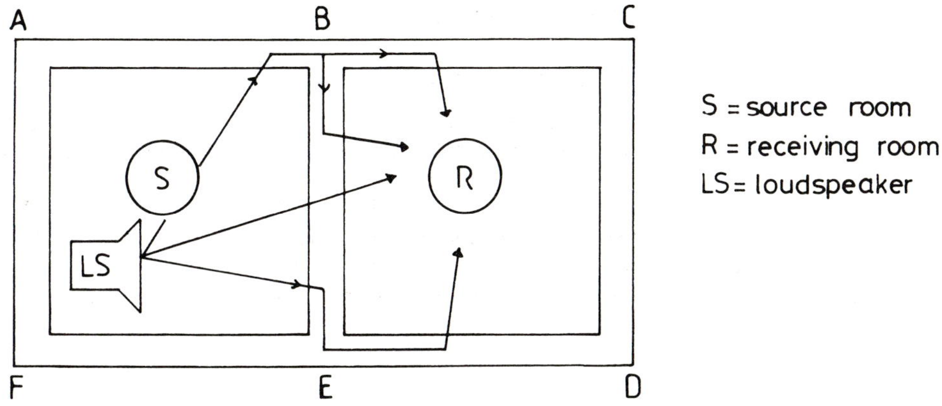

The sound from a loudspeaker in the source room S is transmitted to the receiving room R by a number of paths. Sound waves from the loudspeaker travel through the air and excite the partition wall BE into vibration. The vibration of the wall in turn excites the air in the receiving room. This type of sound transmission is called airborne sound.

Sound waves travelling through the air from the loudspeaker also excite other parts of the structure such as the wall AB. Its vibrations, whether in the form of compressional waves or bending waves, are transferred directly to the section of wall BC. The vibrations of this wall BC excite the air in room R to produce sound. This is called flanking or indirect transmission of airborne sound.

There are other similar paths by which sound gets from S to R, a noteworthy one being that involving direct excitation of the source room surfaces such as might be obtained from the legs of the loudspeaker cabinet, or by the impact of walking or hammering. It is usual to refer to sounds such as these, which originate in the solid structure, as structure-borne sound.

It can be seen that the terms 'airborne' and 'structure-borne' can be misleading. The airborne path involves part of the structure, if only the wall BE, and all the structure-borne paths involve transmission through the air at some stage.

3. THE MASS LAW

If we consider a single-leaf partition (e.g., a single brick wall), the airborne sound transmission is largely governed by the mass law. In the frequency region over which the mass law is obeyed, the sound transmission loss increases by about 5 dB for every doubling of the mass per unit area of the wall (Figure 7.2). It also increases by about 5 dB for each doubling of frequency. In theory, the mass law applies to partitions only if their stiffness can be neglected.

Figure 7.2 Relationship of sound insulation to mass/unit area (from Guidance Note: Sound Insulation, HMSO, 1975)

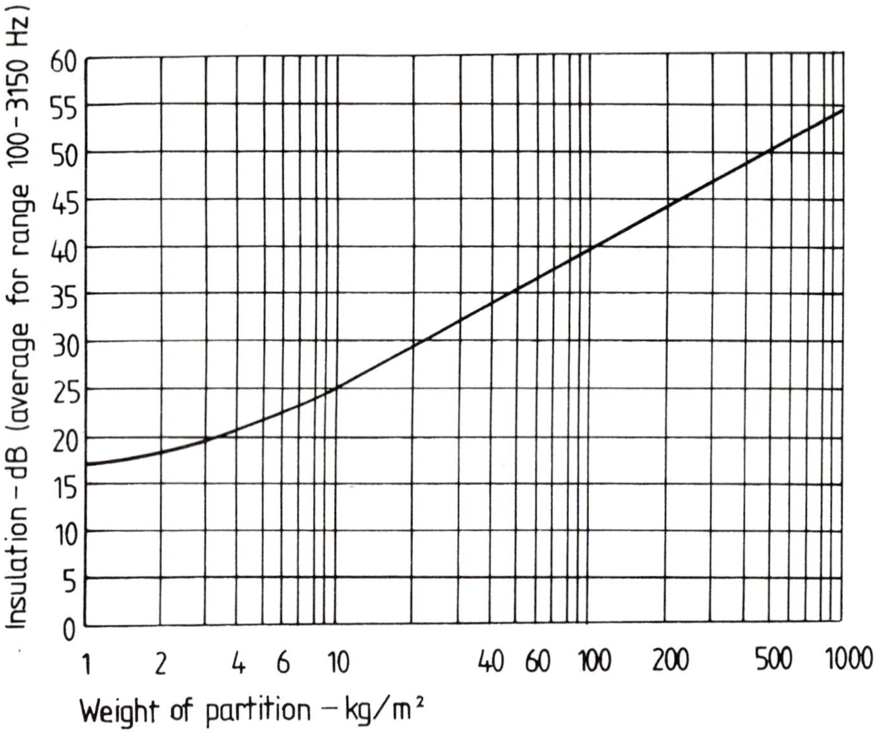

Because of their stiffness, real walls do not obey the mass law over the whole range of audible frequencies [1, 2, 3]. Above a frequency called the 'coincidence frequency', determined by the speed of bending waves in the partition, the insulation falls below the value given by the mass law. The extent to which the insulation falls is not always easy to predict, as it depends on the damping of the partition. The damping of the partition depends partly on the nature of the partition material and partly on the nature of the fixing.

4. DOUBLE WALLS

Because doubling the mass of a single wall gives only a 5 dB increase in sound transmission loss, the law of diminishing returns sets in very rapidly and it is not generally economic to construct a single wall to give a sound-reduction index greater than 55 dB (averaged over the usual frequency range of 100–3150 Hz). Such a wall would have a mass of 1100 kg/m^2 (e.g., 470 mm thick dense concrete).

Where higher degrees of sound insulation are required, or where the structure will not support heavy walls, double walls separated by a substantial air gap must be used. Unfortunately, the sound insulation of two walls is usually much less than the sum of the insulation figures of the individual walls, unless they are more than a few metres apart. This is mainly because the air between the walls acts like a spring, coupling the motion of the leaves together. The stiffness of the spring increases as the air gap diminishes, and with air gaps less than about 30 mm the insulation is about the same as a single leaf having a mass per unit area equal to the sum of the values for the individual walls. Figure 7.3 shows a comparison of the sound insulation of single- and double-wall constructions.

Figure 7.3 Sound-reduction index of single and double walls

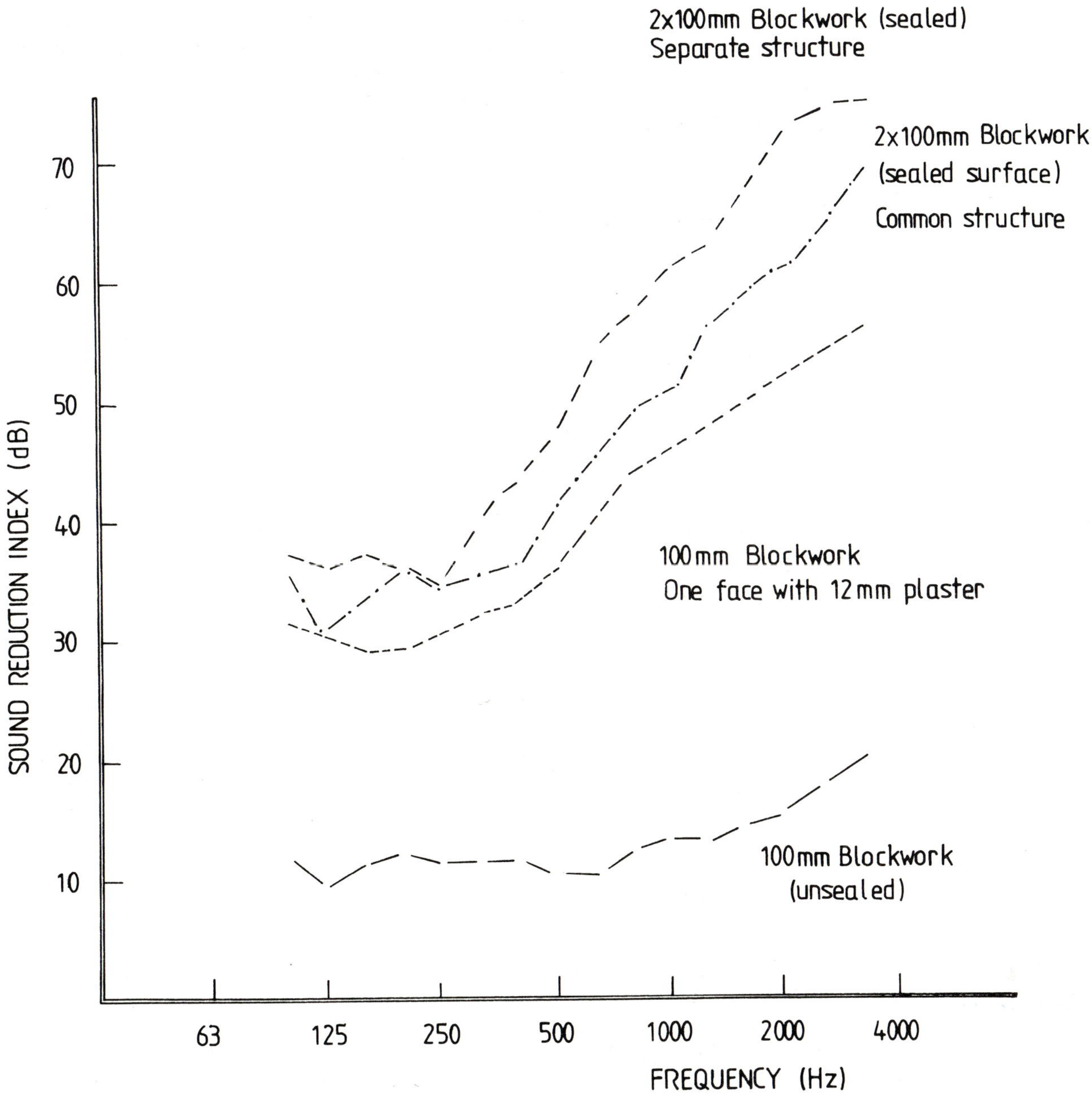

It is well to remember that the stiffness of the air cavity can resonate with the mass of the separate leaves to give a degradation of the sound insulation at certain frequencies. The frequency at which the mass-spring-mass resonance occurs can be readily calculated.

It is possible to calculate the airborne transmission loss of double walls generally, but the errors can be quite significant, especially with large air gaps [1, 2, 3]. High values of airborne sound insulation obtained in the laboratory are often not realised in practical situations, where flanking transmission becomes significant.

5. STRUCTURE-BORNE SOUND

Where high degrees of sound insulation are required, it is necessary to take precautions to ensure that there is no significant sound transmission via the supporting structure of the building. This is often the situation existing when a mechanical plant-room is placed over a noise-sensitive area, such as a conference room. A floating floor resting on resilient supports will usually be installed [1, 2, 3]. It may be necessary to reduce the excitation of the structure via the walls by installing isolated walls as well (Figure 7.4).

Figure 7.4 Floating floor and walls

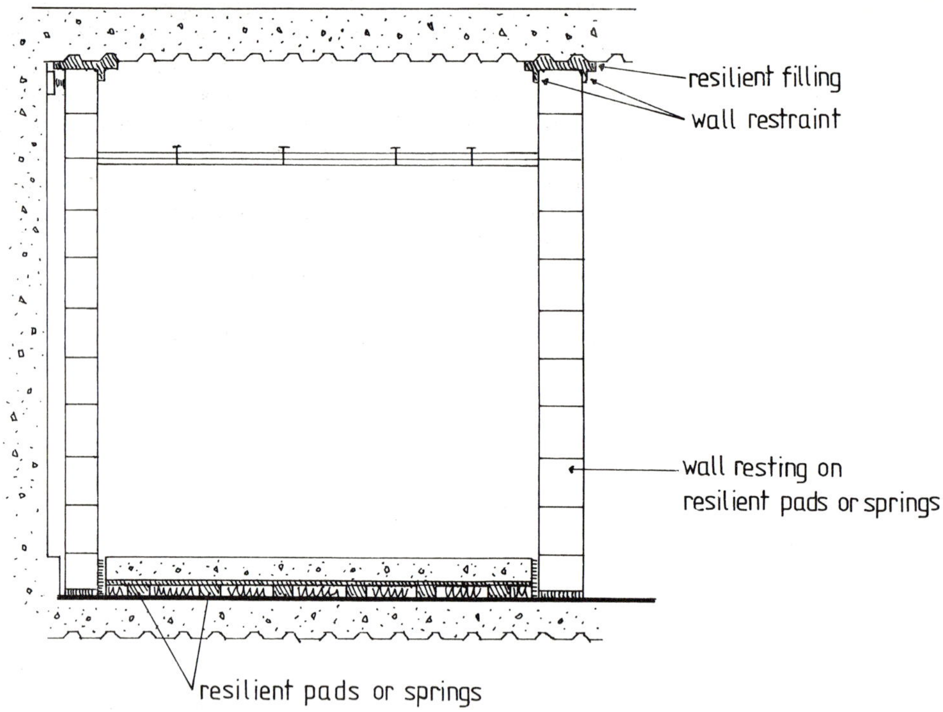

6. DEGRADED SOUND INSULATION

Although the sound insulation of single partitions is determined mainly by their mass and stiffness, there are other factors which can arise in practical constructions which can lead to an unexpected degradation in sound insulation. These are not often reported in the standard textbooks, although the effects have been known for many years. In 1970, Higginson [4] described measurements which showed that expanded polystyrene linings on the external walls of some maisonettes reduced the insulation from Grade I to Grade II. Higginson also reported poor sound insulation in flats where the party walls were of dense concrete with permanent wood-wool shuttering. In both these instances it was suggested that the mass of the internal layer of plaster resonated with the compliance of the resilient lining (polystyrene or wood-wool), thus degrading the insulation. Gösele [5] confirmed the seriousness of the effect when a layer of plaster covers a layer of thermal insulation. Gösele's work on the impairment of sound insulation by the application of certain types of thermal insulation treatment appears to have started as long ago as 1961.

Building Research Establishment papers by Sewell and Utley [6] published in 1975 describe some of the conditions under which filling cavity walls with urea formaldehyde foam can increase the direct and flanking sound transmission.

7. DEVELOPMENTS IN MATERIALS AND COMPONENTS

One of the main developments in the past few years has been the decline in the use of asbestos-based products in the building industry for health reasons. Asbestos was often used as a component in sprayed-on acoustical absorbent treatments of good performance. There are a number of alternative products available which do not contain asbestos, so the restriction is not a serious one for the designer. It should be said, of course, that the acoustical absorbent properties of a material are generally of little relevance to its use as a sound insulator. However, acoustic absorbents are often used in sound-insulating construction as a method of damping out unwanted resonances. Mineral-wool and glass-wool products generally perform this function very well. Open-cell polyurethane foams can also be very efficient absorbers and have been used, although the material will usually need to be of a fire-retardant type.

Generally speaking, there have been no new materials in the past few years which are of great importance for sound-insulation practice. On the periphery of the subject, there are a few developments which it is worth mentioning.

7.1 Sound-absorbent blockwork

For many years there has been a need for an acoustical absorbent which is mechanically hard and able to withstand a moderate degree of mechanical stress. For example, rooms used for indoor ball games are often very noisy inside because the room surfaces have to be hard and durable to withstand the impact of a ball, and nearly all acoustic absorbers are porous and fragile. There has been available, for some years, acoustically absorbent blockwork, an example of which is shown in Figure 7.5. The blocks derive their sound absorption from a slotted construction backed by a closed damped cavity which acts as a damped Helmholtz resonator. The sound absorption can be made greatest at low frequencies but an array of blocks still has useful absorption at middle and high frequencies. The sound insulation of such a wall can increase by the action of the resonators, as shown by Junger [7].

Figure 7.5 Sound-absorbent blockwork (after Junger, 1975)

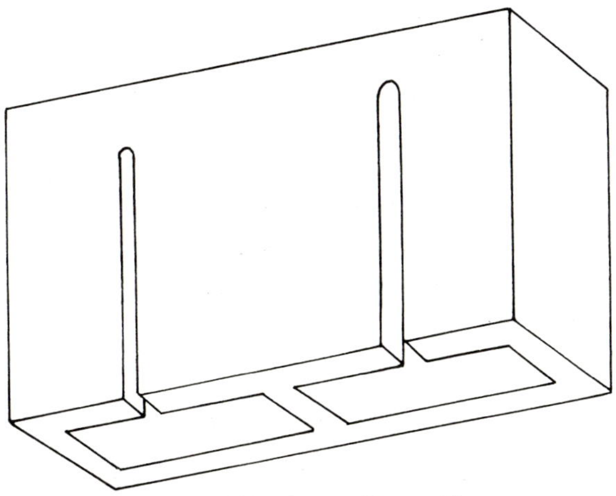

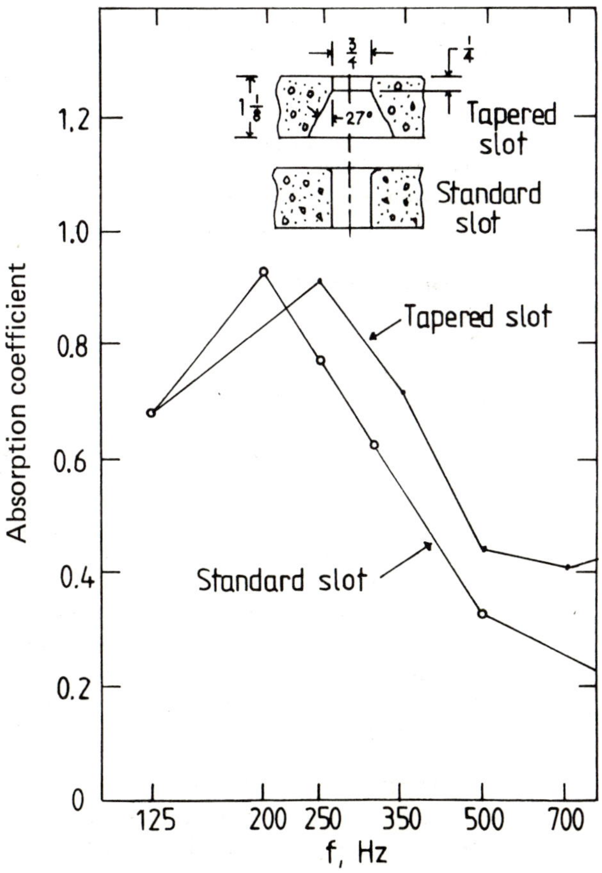

A newer material having possibly the same sort of applications has been developed in the last few years. This material consists of mineral particles bonded together to form a porous rigid-frame sound absorber. It is claimed that the flow resistance, porosity and structure (which determine its acoustic absorption) can be predetermined and adequately controlled in manufacture. The random incidence sound-absorption coefficient is quite good, above 315 Hz (Table 7.1).

Table 7.1 Random incidence sound-absorption coefficient for rigid absorber

Frequency, Hz	*125*	*160*	*200*	*250*	*315*	*400*	*500*	*630*	*800*	*1000*	*2000*	*4000*
Sound-absorption coefficient	.10	.20	.25	.35	.50	.65	.75	.85	.85	.85	.85	.95

The sound-reduction index of a 100 mm thick wall of bricks of the material, plastered on one face, is shown in Table 7.2.

Table 7.2 Sound-reduction index for 100 mm thick rigid absorber, plastered

Frequency, Hz	*125*	*160*	*200*	*250*	*315*	*400*	*500*	*630*	*800*	*1000*	*2000*	*4000*
Sound-reduction index, dB	33	34	39	38	41	41	43	46	49	52	58	64

The material can be moulded to produce blocks, bricks and panels. Its present cost (of the order of £400 per cubic metre) probably limits its widespread application but one can imagine specialised applications where its cost would be justified [8].

7.2 Windows

Windows are usually the weakest path by which external noise penetrates a building and, with the high levels of traffic noise in a modern city, good window insulation is important.

Laminated glass is known to perform better than normal glass. The increased damping raises the transmission loss in the coincidence region. While this has been known for at least ten years, presumably the extra cost has limited the widespread use of this material.

At least one manufacturer of sealed double-glazing units is offering a type containing the gas sulphur hexafluoride, which fills the cavity between the panes. Enhanced transmission loss is claimed for this arrangement at certain frequencies because of the absorption of the gas at the molecular level. Other manufacturers claim that the increase in transmission loss is no more than 2 dB at best and that the additional expense is not justified.

One evolutionary benefit for window design is in the matter of weatherstripping. Bodycombe [9] considers that the standard of weatherstripping has improved so much in recent years that an openable window can be as sound-tight as a fixed one. Bodycombe's paper also discusses the interesting consequences of fire regulations on sound-insulation design and performance.

7.3 High damping polymers

Where very high values of sound insulation are required, a viscoelastic damping layer is sometimes sandwiched between the layers of material forming the sound-insulating partition or floor. The purpose of this layer is to damp out any resonances that the partition or floor might have. This is often done, for example, with the floating concrete floor of a broadcasting studio [10]. Large concrete floors often resonate at quite low frequencies and materials having high damping at low frequencies are hard to find. New polymers under development may be useful if this respect. One example is a urethane polymer which feels like a very tired piece of rubber. Although a solid, it has some of the properties of a very viscous liquid, and has a very low rebound resilience.

8. OTHER DEVELOPMENTS

8.1 Light structures

For simple single-leaf partitions, we know that the sound-insulating properties depend almost entirely on the mass per unit area of the partition and its stiffness. As a rough guide, the sound insulation improves if we increase the mass per unit area of the partition and reduce its stiffness. It is here that we find we are in conflict with the modern structural engineer who is busy devising structures which are lighter but stronger.

A common structure these days is the coffered ceiling. This is a structural member designed to have a high stiffness/weight ratio, the opposite of what the noise-control engineer requires. In trying to predict the insulation of such a structure there are obvious difficulties, such as deciding what we take as the mass per unit area of such a non-uniform element.

Modern developments in lightweight structures, such as tensioned-skin and inflatable enclosures, may well be limited by deficiencies in acoustic performance. Not only do such structures present severe problems in sound-insulation design, but they also often have the kinds of shape which produce severe internal focussing effects.

8.2 Microprocessors

There seem to be two ways in which the microprocessor is likely to have an effect on the design of buildings, so far as acoustics is concerned. One is in the provision of inexpensive computing power; the other is in the application of electronics to the control of noise, so-called 'active attenuation'.

8.3 Microcomputers

For under £3000 one can now buy, off the shelf, a computer which is more powerful than those available ten years ago at three times the present price. The full implications of this are yet to be developed. Until recently, the acoustic designer of buildings was frequently in the situation where he had neither the time nor the resources to carry out a detailed analysis of a proposed design. With the present availability of relatively cheap microcomputers and minicomputers, certain acoustical calculations can be carried out much more efficiently. In the author's practice, several acoustical programs have been developed over the past six years or so. Before microcomputers were available, the firm subscribed to a timesharing computer bureau and, although the bureau was very efficient, only two programs were run very frequently. One of these calculated the reverberation time as a function of frequency for a room, and allowed the effects of changes to the quantities and types of absorbent to be quickly evaluated. The other program calculated, sorted and classified all the modal frequencies for a rectangular room. These programs were dominant because it was foreseen that the time spent in developing the programs would be well repaid by the time saved in using them. Many other potential programs (or even commercially available programs) were not developed or used, because they were not cost-effective. An example of this was a program for calculating the silencer attenuation required in a ventilation duct system. With such a program, the time spent connecting to the bureau and typing in the basic data was excessive compared with the computation time; such calculations were done more effectively using a simple programmable calculator.

Now that the bureau service has been replaced by an in-house microcomputer, many more programs, previously thought to be uneconomic, have been developed. It is not just the microprocessor that has changed the situation, but also the inexpensive storage of data, which is, perhaps, even more important. An example of the influence of this factor is in the computation of cross-talk via ventilation ducts between rooms. In order to calculate the cross-talk attenuation required between any two rooms so that the sound insulation is not degraded, one needs to input the duct length, cross-sectional areas of ducts and branches, and receiving room absorption. With a complex of a dozen or so rooms, the calculation takes about eight hours with a programmable calculator. As most of this time is spent reading dimensions from a drawing and keying-in the data, this does not, at first sight, seem a worthwhile job for a computer. However, the benefit of computer data storage becomes apparent when, as so often happens, the calculations have to be repeated because the design has changed in the interim. With the original input data stored on floppy disk, it is a comparatively simple matter to re-run the program, with only a few keystrokes to modify the altered input parameters. Moreover, the acoustic designer can thereby provide a rapid response to the changed conditions, having done most of the work in the first keying-in of data.

There are indications that it might well be possible in the future to calculate sound transmission in building structures more reliably by the use of statistical energy analysis

[11]. This technique requires further development before it is more widely used, but the availability of inexpensive computing power should accelerate the process.

9. ACTIVE ATTENUATION

Active attenuation is an interesting area in which microprocessors have a likely application. The basic concept is quite old and quite simply stated. In order to attenuate a sound, one simply generates 'anti-sound' with a loudspeaker in such a way that it cancels out the unwanted sound. The concept is simple but the practical realisation is so difficult that it has eluded researchers for over thirty years.

Within the lost ten years or so, successful anti-noise devices have been built for simple, almost one-dimensional, systems like ventilation ducting, but the most impressive development has been for a system which suppresses low-frequency noise at the source and is effective for all the surrounding area [12]. The present restriction with the system is that it relies on the noise being periodic so that a synchronising signal is available.

Figure 7.6 is a schematic illustration of the device applied to the exhaust of a diesel engine. The microprocessor is programmed to generate synthetically an electrical waveform which drives the loudspeaker. The waveform is altered by the microprocessor and if the alteration produces a lower sound output as sensed by the microphone, this waveform is stored until a better one is found. This trial and error process is continually repeated. After about twenty seconds a good 'anti-noise' waveform is produced and the environment much quietened. Since the criterion is the sound level at the microphone, the system works over a wide area only for long-wavelength (i.e., low-frequency) sounds. However, this is not a serious disadvantage, because middle- and high-frequency noises are comparatively easy to attenuate by conventional methods. Until quite recently, excessive low-frequency noise has often been an insoluble problem.

Figure 7.6 Cancellation of repetitive noise

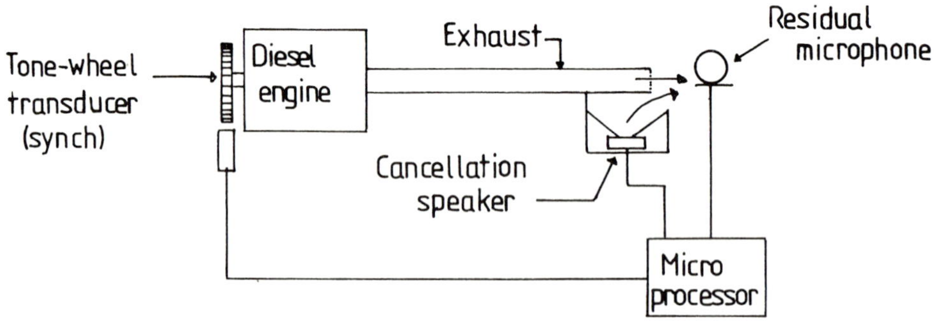

The work is being extended by attacking the problem of structure-excited noise and vibration. For this, the loudspeaker is replaced by an electrodynamic vibrator attached to the structure and the microphone is replaced by an accelerometer.

REFERENCES

1. Parkin, P.H., Humphreys, H.R. and Cowell, J.R. *Acoustics, noise and buildings* (4th edn), Faber and Faber (1979).
2. Croome, D. J. *Noise, buildings and people*, Pergamon Press (1977).
3. Harris, C.M. (Ed.) *Handbook of noise control*, McGraw Hill (1979).
4. Higginson, R.F. 'Sound insulation between rooms having resilient linings on the walls', *Applied Acoustics*, (3), 133–43 (1970).
5. Gösele, K. 'Schallschutz von Bauteilen aus Beton', *Beton*, **26**, (1), 26–30 (1976).
6. Sewell, E.C. and Utley, W.A. *The effect of resilient fillings on direct and flanking sound transmission with cavity masonry walls*, Building Research Establishment, Current Paper CP10/75 (January 1975).
7. Junger, M.C. 'Helmholtz resonators in load bearing walls', *Noise Control Engineering*, **4**, (1) (January/February 1975).
8. I.C.I. *Rigid acoustic absorber DS5269*, Brochure available from Bestobell Acoustics, Slough, England.

9. Bodycombe, H. 'Acoustics vs fire precautions in offices', Proceedings of the Institute of Acoustics, Paper 1.15, 73–6, Spring Conference (1980).
10. Ward, F.L. and Randell, K.E. 'Investigation of the sound isolation of concrete slab floors', *J. Sound Vib.*, **3**, (2), 205–15 (1966).
11. Craik, R.J.M. 'Sound transmission through buildings using statistical energy analysis', Proceedings of the Institute of Acoustics, Paper 1.13, 65–7, Spring Conference (1980).
12. Chaplin, G.B.B., Smith, R.A. and Powell, A.R. *An adaptive waveform generator for cancelling engine exhaust noise*, Institution of Electronic and Radio Engineers, Colloquium on Acoustic Noise Cancellation (published in January 1980).

8 Design aids and criteria for sound insulation

Robin Mackenzie

SYNOPSIS

The implementation of acoustic design criteria in the building industry entered a new phase following the publication of the CIBS guide *Environmental criteria for design* [1]. This guide helped to consolidate existing criteria and to add a useful interpretation in an area of applied science which for many years has frequently been misinterpreted by architects and engineers alike. The second phase, however, will prove much more difficult to achieve, since it will involve a long-term process of education to ensure that the criteria are adhered to and implemented correctly. Equipment manufacturers will have to be persuaded to supply sound-power and vibration-level data according to frequency bands, as a matter of standard practice. At present, manufacturers show a wide variation in their ability to supply data, without which design criteria are meaningless.

Education of both professional designers and supervisory staff in the practice of noise and vibration control is limited and should be reviewed by the relevant professional bodies.

1. INTRODUCTION

During the last five years we have seen in this country a very considerable growth in public awareness of the problem of noise.

In many cases these problems relate to environmental or industrial noise, that is, types of noise which can be controlled by standards and guidelines laid down by the Control of Pollution Act 1974.

Noise within buildings does, however, tend to be of a different nature and its effects and effective control are accordingly somewhat more complex.

In dwellings, standards for party walls and floor insulation, together with 'deemed to satisfy' constructions, are provided for within the Building Regulations. For other buildings, however, there are no such provisions within the Regulations and the only criteria available are those contained within the guides produced by the CIBS and RIBA, respectively.

In practice, two factors have contributed to the growing problem of noise in buildings in recent years. Firstly, there has been increasing use of lightweight partitions, which, in general, have less acoustic impedence and greater resonant properties than the more traditional brick or blockwork constructions. Secondly, the demand for higher standards of environmental comfort through the provision of fully air-conditioned spaces has, in many cases, given rise to excessive levels of ambient noise, by virtue of inadequate design.

1.1 The effects of noise

The effects of noise on people have been extensively studied, classified and, to some degree, quantified. In the main, the effect of audible-acoustical energy on people falls into four general overlapping categories, as follows:

1. Demonstrable hearing loss, accompanied by the social ramifications of that loss
2. Interference with the ability to communicate or to hear desired sounds or acoustical signals

3. Annoyance and irritation effects of varying degrees, such as interference with work or sleep
4. Physiological reactions.

The most obvious effects of noise on people are auditory. One set of auditory effects is noticeable after a noise has disappeared: this consists of temporary hearing loss, permanent hearing loss and permanent injury to the inner ear. Another set of auditory effects is noticeable while a noise is present: this consists of masking and interference with speech communication. Both sets of auditory effects are adverse in terms of human response.

For the purposes of this chapter, the problem of high-level noise will be disregarded, since the means of control in order to prevent damage to hearing is relatively straightforward and well defined within the *Code of Practice for reducing the exposure to noise of employed persons at work* [2].

Noise at a much lower level can, however, interfere with the perception of audible signals. This is called masking. By masking, an auditory signal can be made inaudible or the signal can be changed in quality and apparent location. The facts of auditory masking are well established, and the masking effects of noise can often be calculated from measurements of the signal and the noise.

An important instance of masking is the interference with speech communication that results from noise. Figure 8.1 summarises the relations between interfering noise and the possibilities for speech communication. The vertical axis is the A-weighted sound level of the interfering noise, while the horizontal axis is the distance in feet between the talker and listener. The area near the bottom of the graph (the lightly hatched region below the heavy curved line) represents the combinations of distance and levels of interfering noise for which speech communications can be nearly normal. Speech communication situations involving groups or pairs of individuals often involve speaker-listener distances of 5 to 12 feet, corresponding to levels (for interfering noises) of 66 to 55 dB(A).

In a highly intellectual, technical society, speech communication plays an extremely important role. Noise can reduce the accuracy, frequency and quality of verbal exchange. In excessive noise formal education in schools, occupational efficiency, family life, the

Figure 8.1 Speech communication, with distance

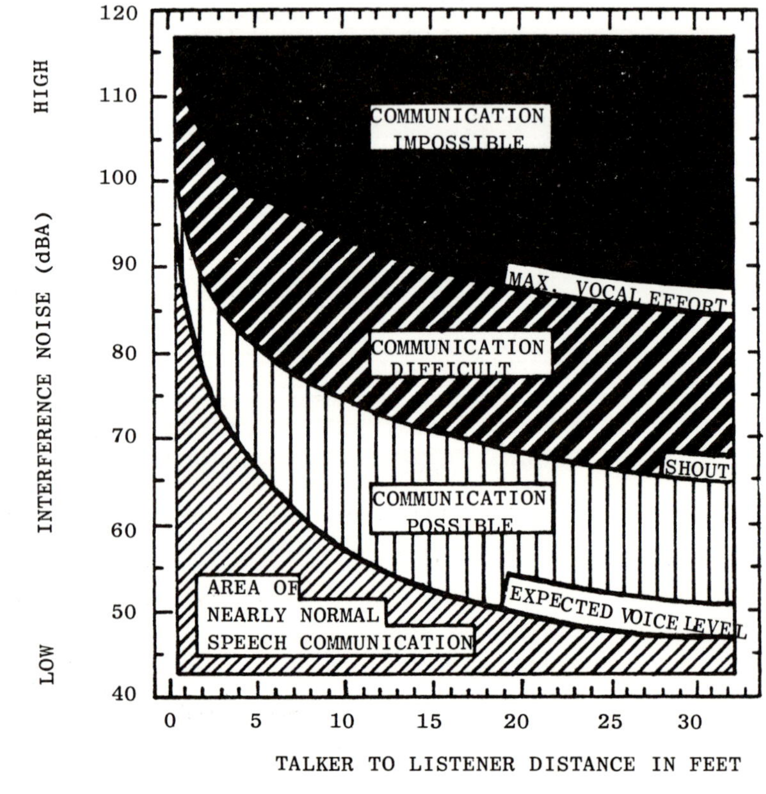

quality of relaxation and the enjoyment of life can all be adversely affected. Speech reception by elderly persons seems to be especially affected by noise.

Noise not only has direct auditory effects but also produces psychological and sociological effects of a more general nature. If a task requires the use of auditory signals, either speech or non-speech, then noise at any level sufficient to mark or interfere with the perception of those signals will interfere with the performance of the task. When mental or motor tasks do not involve auditory signals, steady noises without special meaning do not seem to interfere with the performance of these tasks, unless the noise level exceeds about 90 dB(A). Even above these levels, performance is somewhat unaffected. On the other hand, irregular, unpredictable bursts of noise may influence performance when their noise levels are less than 90 dB(A).

It has also been shown that the effect of noise upon the number of errors committed varies with the conditions of work and the state of the subject. Noise heightens arousal; hence, if people are short of sleep and are doing undemanding work of a routine nature, noise may, by arousing and stimulating them, actually diminish the number of errors made. On the other hand, if people are doing work which, by its nature, already maintains a state of alertness, a loud noise may cause them to become jittery and make more errors. This result supports the general experience that routine work is less affected by a loud-noise environment than is work of an exacting kind.

However, recent research [3] in the effects of ventilation fan noise upon clerical efficiency showed that the fan noise had no effect upon either output or error rate. However, it was found that background speech reduced output by 13 per cent. In other words, the degree of coherence or information level of the background noise is important in the degree of distraction. This in fact was verified by further research using speech with varying interest levels.

The findings of this research more or less agree with the well-established Yerkes-Dodson Law, which can be interpreted to show that noise can be beneficial and act as a stimulant for simple tasks which easily become boring and have little motivation, whereas for complex tasks noise is detrimental.

With regard to the more serious psychological effects of noise, there is some evidence that admissions to psychiatric hospitals are higher in areas with high noise levels than in quieter areas but such evidence is not entirely convincing. There is no evidence that exposure to noise can result in mental illness. However, all of the facts clearly support the contention that noise can be a source of psychological distress through annoyance, disturbance of activities such as sleep and speech communication, and so on. Psychological distress, in turn, can contribute to a list of symptoms such as nausea, irritability, general anxiety and changes in mood.

Evidence from a joint Polish/American study [4] has shown that disorders of a mental nature are likely to be higher in cases where exposure is higher, whereas physiological disorders are unaffected. Although there is little evidence to suggest that low-level continuous noise is likely to produce a harmful physiological reaction, the effects of transient noise are well established and are given as follows:

(a) Fast response of the voluntary musculature that are mediated by the somatic nervous system

(b) Medium response of the smooth muscles and glands that are mediated by the visceral nervous system

(c) Slow response of the neuro-endocrine system.

Particular attention has been given of late to the study of the neuro-endocrine responses, which produce effects similar to the responses to stress.

The response to stress, called the general adaptation syndrome, consists of three stages: an alarm reaction, a stage of resistance and a stage of exhaustion. If a stress is severe and is maintained for prolonged periods of time, an organism passes in succession through the stages of alarm reaction, of resistance and of exhaustion. In the extreme case, the end result is a breakdown of bodily function and death. Even in the less severe case, a price may be paid for continued stress during a prolonged stage of resistance. This price may include increased susceptibility to infection and, perhaps, specific diseases known as the diseases of adaptation. Such diseases may include, among others, some types of gastro-intestinal ulcers, some types of high blood-pressure and some types of arthritis. It should be said, however, that many medical authorities do not accept the theory that there are diseases of adaptation. Rather, they theorise that each disease has its own special set of causes.

As a final word on the effects of noise on health and performance, one should make reference to the sociological impact of noise. For example, due to ethnic background, one group of workers may accept a noise environment in their office that would be considered unacceptable by those of a different cultural orientation. They may, in fact, create conditions that, while acceptable to themselves, are considered noisy by others.

This phenomenon must be taken into account in assessing the attributes of noise as a sociological problem. It also must be given careful attention in translating results of studies on noise as related to a particular source and affecting a specific population.

1.2 Design procedure

The design procedure for sound insulation in buildings involves three basic steps, as follows:

(a) Establishing the desired noise rating for a particular space
(b) Determining the sound level of adjacent sources of noise
(c) Calculating the required noise reduction and selecting a suitable form of insulation.

These steps will now be considered in turn, giving details of sources of data, together with problems likely to be encountered in the design process.

(i) Noise rating

In an idealised design procedure, the required sound-pressure level within each octave band will be established for a particular situation. Such a method is, however, rarely

Figure 8.2 NR curves

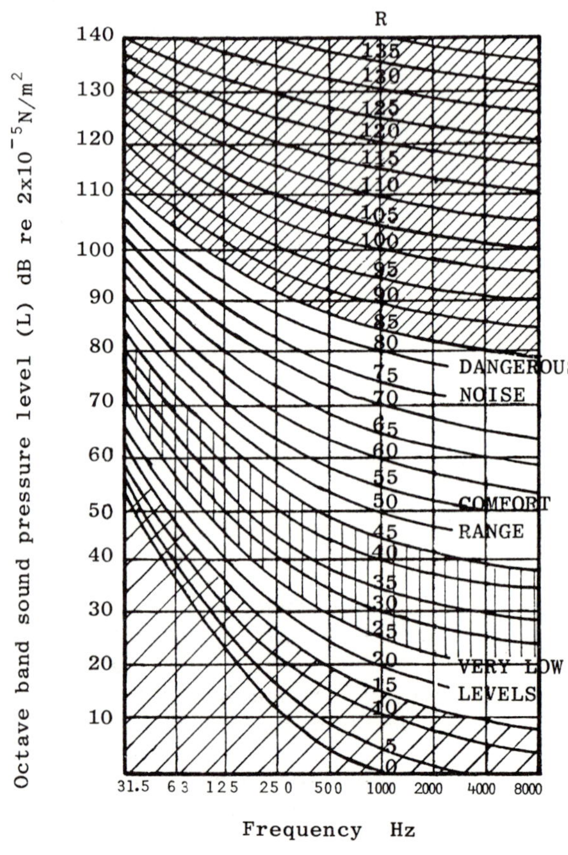

employed except for specific design exercises involving stringent acoustic requirements, as would be the case with concert halls or similar buildings.

For routine design work, noise rating criteria in the form of NC or NR curves (see Figure 8.2) are commonly used, although a specification may simply relate to a level in terms of dB(A). These curves have been developed to take account of subjective response to sound in terms of frequency and for most applications will be perfectly adequate. There is only a slight difference between the values of the NC and NR, the former being used more commonly in North America, while the latter has been widely adopted in Europe as the standard criterion. To all intents and purposes, the values may be interchanged without affecting the overall results.

The use of dB(A) is more rigid, in so far as it applies a spectral correction according to a fixed set of values, unlike the NC/NR curves, which vary in terms of weighting according to level. For this reason, the NC/NR curves are to be preferred.

Despite the reasonable success in using NC/NR curves, acoustic consultants have become increasingly concerned about the low-frequency rumble associated with buildings designed using these curves. The bias adopted in the NC/NR curves borrows heavily from the concept of equal loudness, which does not necessarily equate with equal annoyance. Indeed, in designs involving large centrifugal fans, boilers or other units having predominantly low-frequency sound output, consultants will tend to use the Preferred Noise Criteria curves (see Figure 8.3), which are more demanding below 250 Hz than the other criteria.

Figure 8.3 PNC curves

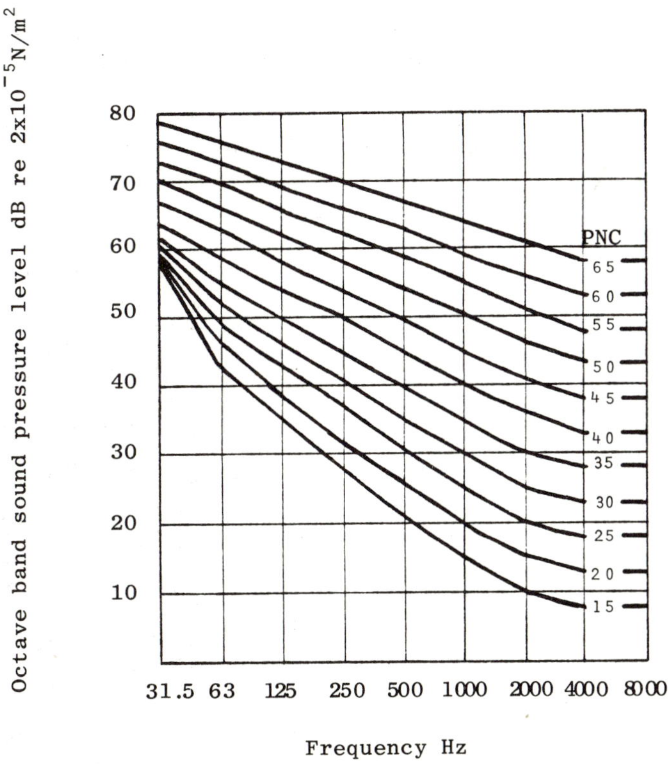

In cases where the source has some special spectral or temporal characteristic, for example, discrete tones or impulsive noise, then a correction will have to be applied to the criteria. Normally, this will simply result in a more demanding curve being recommended (–5 units is commonly applied), although the exact value of the correction should be varied depending upon the exact nature of the source.

Psychological effects will also enter into the calculation in so far as, with some sources of noise, the only acceptable criterion is one of virtual inaudibility. Examples of such noises include that arising from the use of toilets or the transmission of private conversation or, alternatively, from locations where foul language may be overheard.

A less obvious factor involves the relationship between the source and receiver. A person will have a greater tolerance to intrusive noise from a source to which he can relate. For example, whereas a computer programmer may be able to work quite effectively in an office with a background level of NR 60 resulting from computer printing machines, the same person is likely to find such a level intolerable if the source is a conveyer belt or other item of machinery unrelated to his own work.

While provision can and should be made for these general psychological factors, individual response, including behavioural characteristics and personality, cannot normally be catered for and, as such, can only be treated on a general basis.

(ii) Source levels

In order to achieve an efficient design for sound insulation, the source noise level should at least be specified in terms of octave-band sound-power levels. Such a band width may, of course, fail to identify the presence of discrete total characteristics and, therefore, if one is unfamiliar with the equipment, the manufacturer should be asked to state whether or not such tones are produced. Similarly, a check can also be made at the point with regard to any impulsive or intermittent characteristics.

Should the only information be in the form of a sound level in dB(A), one is unable to optimise the sound insulation required at different frequencies and this inevitably will result in a more expensive design being used than would be the case if spectral information had been provided.

One aspect which often leads to unnecessary confusion is the inconsistent use of Sound Power Level (L_W re. 10^{-12} watts) and Sound Pressure Level (L_p re. 2×10^{-5} Pa*). Mistaking one set of values for the other can have serious consequences, either in terms of performance or cost, or both.

The Sound Power Level is an acoustic source power and is considered to be independent of the environment in which the source is situated. Sound pressure is affected by the room volume and absorption and also the distance between the point of measurement and the source.

The Sound Power Level of a machine will normally have been measured in an anechoic chamber or well defined acoustic environment, and carried out according to an appropriate standard such as BS 4196: 1967 [5]. The Sound Pressure Level may also have been measured according to a standard method under laboratory conditions but frequently is

Figure 8.4 Predicting sound level in semi-reverberant fields

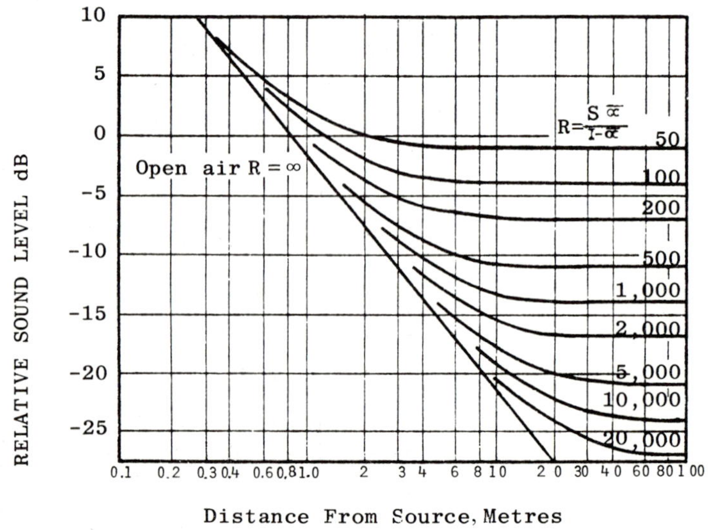

* Note pascals are equivalent to N/m^2 units.

the result of a field test. Under the latter circumstances, the absorption characteristics of the room in which the measurement has been carried out must be considered and the room constant R applied as shown in Figure 8.4.

$$R = \frac{S\bar{a}}{1 - \bar{a}}$$

where R = room constant in square metres
S = the total surface area of the room in square metres
$\bar{a}$ = average absorption coefficient.

Under reverberant conditions in rooms with hard reflectant surfaces, the Sound Power Level may be related to the Sound Pressure Level by the following equation:

$$L_W = L_p + 10\,\lg R - 6\text{ dB}$$

(iii) Wall selection

Once the information with respect to source and receiving room has been obtained, the calculation of Noise Reduction (NR) required can be simply obtained by subtraction. It should be noted at this point that the NR is not equivalent to the Transmission Loss (TL) or Sound Reduction Index (SRI), since the latter two terms refer to a level difference which has been normalised for receiving room absorption. These terms are frequently misunderstood and misinterpreted, and the correct relationship is defined as follows:

$$TL = SRI = NR + 10\,\lg\,(R + 4S_w) - 10\,\lg\,4R$$

where S_w = surface area of the wall in square metres.

When the required Noise Reduction has been established, reference will have to be made to technical specifications of proprietary wall types. It is very comon for block or partition manufacturers to quote only a single-figure average insulation value in their specification. This may have been done for the sake of brevity but it may also have been used to cover up a serious deficiency at some particular frequency. A wall should *never* be selected on the basis of an average insulation value, especially since the average figure must have been derived in the first place from a frequency-band analysis. The minimum requirement should be for octave band values and preferably one-third octave values normally for the frequency range 125–4000 Hz, although a wider range may be necessary, depending on the source spectrum.

It is also necessary to establish the surface finish to the wall for which the data are supplied. The sound transmission loss of open-textured concrete blockwork is substantially influenced by whether or not a surface finish has been applied [6] (see Figure 8.5).

Figure 8.5 Comparison of effect of different surface finishes upon transmission loss

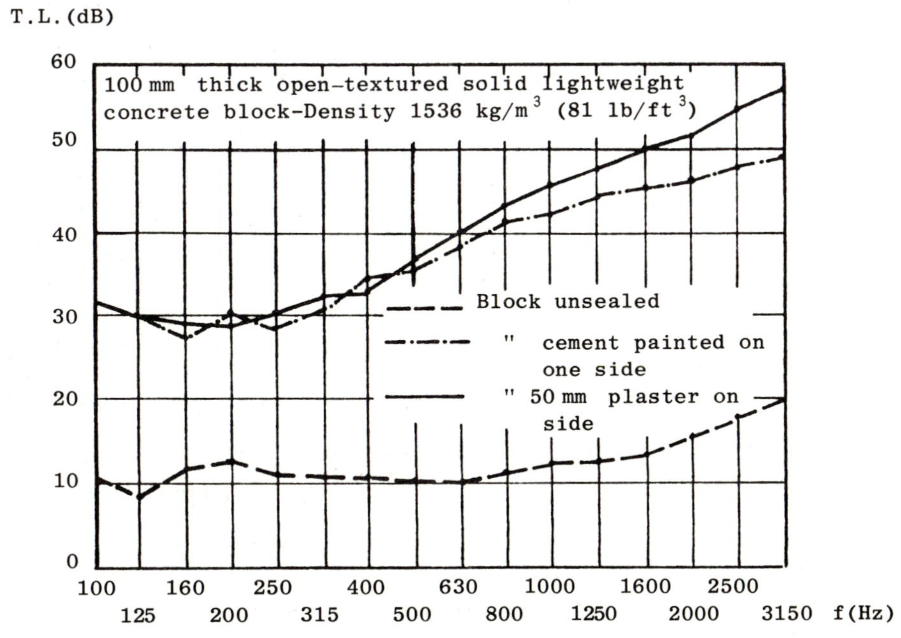

In converting the information given in technical specifications, which is normally quoted in the form of transmission-loss values, to the Noise Reduction values required for the design exercise, it is necessary to establish the receiving room absorption coefficient. This exercise involves a straightforward calculation including a multiplication of the area of each surface by the absorption coefficient. The problems which may arise here lie once again in the accurate provision of information in materials specifications. For example, the absorption coefficient of a suspended acoustic-tile ceiling is very different from that of an identical tile finish applied directly to the soffit of a concrete slab. It is very important to establish the conditions under which the specified data were obtained and to compare like with like.

Similarly, the transmission-loss data are often obtained under laboratory conditions in the absence of flanking paths, which may contribute significantly to the transmission in the field. This aspect will be considered in greater detail in the next section.

2. DESIGN LIMITATIONS – ARCHITECTURAL

One of the major problems in designing for sound insulation is that it is almost impossible for an accurate value to be calculated from basic theory. Unlike more predictable areas of design, such as structural strength, the resultant sound insulation of a wall or floor is very much influenced by three factors, namely, resonance, flanking paths and workmanship.

Traditional education at colleges for both architecture and building-services students has focussed heavily on the 'mass-law', with or without empirical corrections, as a design guide. Thus, the transmission loss is given by the mass law:

$$TL = 20\lg\left\{\frac{m\,\omega}{2\rho_0\,c_0}\right\}$$

where m = surface density of the material (kg/m^2)
ω = angular velocity = $2\pi f$ (rad/sec)
ρ_0 = density of air (kg/m^3)
c_0 = velocity of sound in air (m/sec)
f = frequency in Hz.

Use of this equation, even with empirical corrections, can be very misleading even in the absence of flanking paths and with ideal workmanship.

For most practical constructions including both masonry and lightweight partitions, the insulation is controlled by panel or cavity resonances or a complex combination of both, and not, as generally thought, by mass impedence alone. Theory is now available [7] to allow more accurate predictions to be made but this is somewhat complex, as shown in the equation below:

$$TL = 20\lg\left\{\frac{m\,\omega}{2\rho_0\,c_0}\right\} + 10\lg\left\{\frac{2}{\Pi}\,\frac{f^{1/2}}{f_c\,\sigma^2}\right\} + 10\lg\,(\eta f^{1/2})$$

where f_c = critical frequency (Hz)
σ = radiation efficiency (dimensionless)
η = total loss factor (dimensionless).

For practical purposes, however, this equation may be simplified and represented in graphical form as shown in Figures 8.6 (a) and (b).

While greater accuracy may now be used in predicting the transmission via the direct path, there still remains the problem of flanking transmission via the floor slab or adjacent wall.

Once again, theory is now available [7] which will allow reasonably accurate prediction of sound via indirect paths but the procedure is complex and cannot be readily simplified.

Some general assumptions can be made, however, in so far as common building materials are concerned. The incidence of flanking transmission is increased by the application of lightweight panelling to the face of the flanking wall. This creates a resonant layer

Figure 8.6 (a) Relationship between critical frequency and longitudinal wave velocity for common building materials

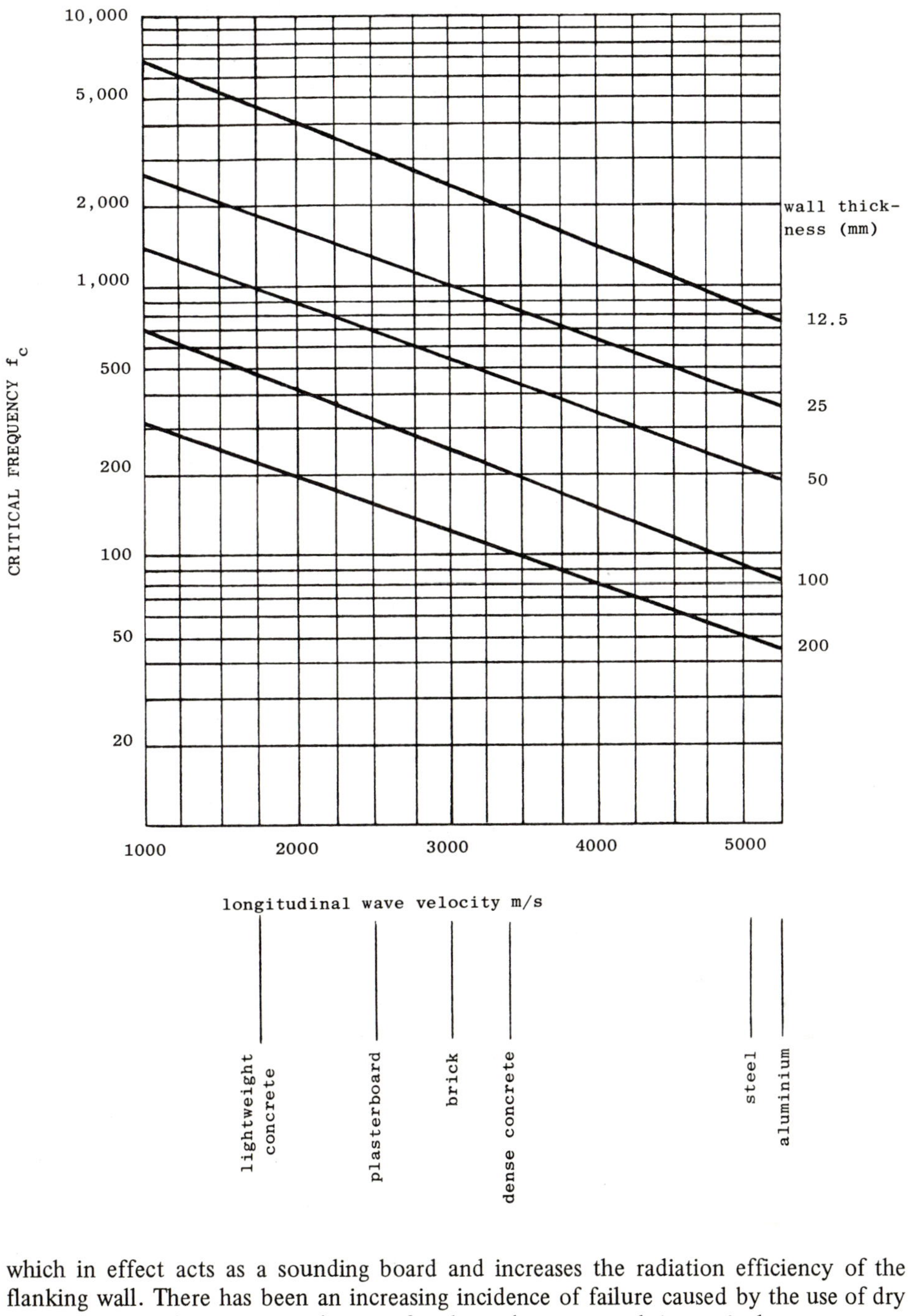

which in effect acts as a sounding board and increases the radiation efficiency of the flanking wall. There has been an increasing incidence of failure caused by the use of dry lining, in part, due to its application for thermal reasons and, in particular, as a means of reducing condensation.

Indeed, it is perhaps slightly ironic that two of our main problems in housing today, that is, noise transmission and condensation, have been exacerbated by one basic design change in the home, that is, the removal of the chimney breast.

The chimney breast added mass and stiffness to the party wall, certainly in the case of brick walls, and gave a design that rarely failed to meet the standard. A 250 mm brick wall today, without such bracing, is a much more marginal form of construction in terms of sound insulation.

Recent changes in building involving the use of new materials or techniques can sometimes have adverse effects on sound insulation between dwellings.

Figure 8.6 (b) Transmission loss as derived from critical frequency of material

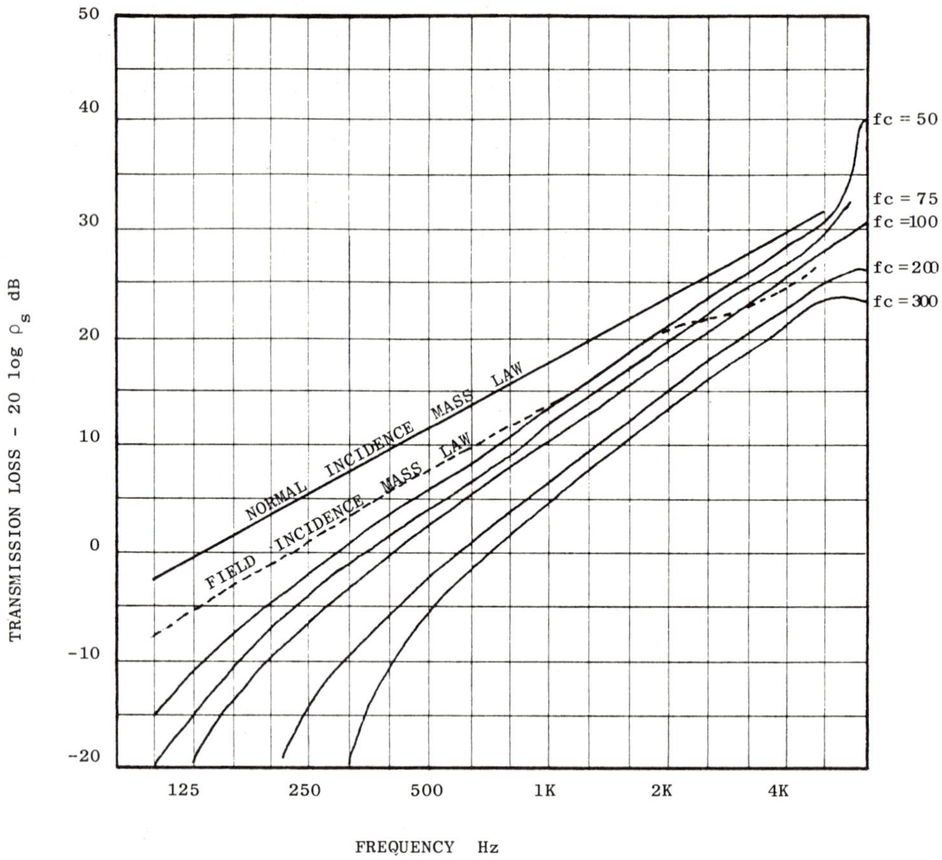

Two particular examples involve the use of urea formaldehyde cavity-wall infill to increase thermal insulation. Sewell and Utley [8] at the BRE have shown that cavity filling with fairly stiff foams can increase both direct and flanking sound transmission with masonry walls. Field measurements have shown large reductions in sound insulation about a week after filling, and smaller, but still substantial, reductions after about a year.

Rigid adherence to the basic concept of mass law has given rise to problems in other ways. For example, it is not uncommon for designers to presuppose that a deficient brick or block wall may be upgraded by the application of additional mass in the form of dry lining with plasterboard. It is in such an application that a paradox may arise, whereby the insulation may actually become worse at lower frequencies after applying the lining, if due care is not taken. The application of plasterboard to battens fixed directly to a wall involves the addition of a highly resonant cavity-panel system to the existing wall, with possibly a higher radiation efficiency as a result.

Care must be taken in the application of such lining to ensure that the cavity resonances are damped either by the insertion of mineral-wool quilt between the battens, or, alternatively, by erecting a free-standing framework upon which to attach the plasterboard and so reduce the panel resonance by effecting a structural discontinuity.

Another misapplication of the mass law is found in the application of mineral-wool infill. When fibrous material is used to dampen cavity resonances in lightweight stud partitions, it is sufficient only to use a low-density infill, say 2.4 kg/m^2. The use of denser materials will only provide marginal increases and will be less cost-effective than, say, additional layers of panelling.

In terms of cost-effectiveness, stud partitions can provide an attractive alternative to brick or concrete block walls, but greater care must be adopted in the design and execution of the work. Small air gaps, less than one mm wide, at structural junctions between the bottom or top rail of a stud partition and the floor or ceiling can seriously reduce the insulation value of a partition. Skirtings and cover plates which are aesthetically quite suitable are frequently found to cover up inadequate edge sealing. Walls must be airtight to be acoustically effective and inevitably this means greater supervision of work in the case of stud partitions, and/or more detailed specifications.

The aspect of low resistance to air flow in suspended ceilings is a frequent cause of flanking transmission at the wall head, in designs where the partition is only built to the underside of the ceiling and not taken up to the structural slab. Many suspended ceilings are either porous or perforated in order to increase the *absorption* qualities; the *insulation* properties are correspondingly reduced, resulting in transmission-loss values as low as 3 or 4 dB(A) in some cases and certainly below 10 dB(A) in many proprietary systems.

There is still considerable confusion amongst designers and contractors over the meaning of sound absorption and sound insulation. Unfortunately, it is still common practice to find absorptive tiles being specified as surface finishes to walls and ceilings in order to increase transmission loss.

Another area of misapplication lies in the specification of double-glazing systems for sound insulation. Here, there is a mistaken belief by some designers that elements providing good thermal insulation will also provide good sound insulation. The building scientist will, of course, recognise that there is no basis in physics to draw such an analogy, indeed, the underlying laws governing heat and sound transfer are fundamentally quite different. Nevertheless, it is still not uncommon to find hermetically-sealed glazing units being supplied for insulation against traffic noise. The width of the gap is, of course, the dominant parameter and although the optimum width is a function of frequency, for most traffic noise applications a gap of between 100 mm and 200 mm is desirable [9].

3. DESIGN LIMITATION – SERVICES

Probably, the most commonly occurring problem today facing the asoustic consultant is how to prevent the transmission of noise and vibration from the plant-room to adjoining spaces; cross-talk and mechanical transmission via pipes and ducts are common. This problem is exacerbated in many buildings by the architects' and engineers' preference for roof-level plant-rooms, generally for perfectly good reasons. Nevertheless, it is a fact of life that the top floor of hotels or office blocks tends also to be the most exclusive part of the building and, as such, often requires a very low noise rating. The architect should be fully aware of the additional cost penalty of placing plant-rooms next to executive floors. The massive increase in slab thickness, the resultant increase in the structure to carry this load, the cost of sophisticated anti-vibration mountings or floating slab design, or resiliently hung ceiling, all add up to a considerable additional sum of money.

One of the major headaches in such design work is the frequent lack of information concerning the sound-power levels of air-conditioning plant and, worse still, the total absence of any information concerning vibration levels at the mounting points of such equipment. The lack of such information leads, by necessity, to excessive 'safety' factors being used, with resultant cost implications.

Although the 'basic' elements of noise and vibration reduction procedures are well documented in publications, frequent faults still occur due to the fixing or suspending of noise-bearing services without resilient brackets or hangers. Many ventilation systems still produce excessive noise at the outlet, owing to regeneration as a result of air-flow turbulence.

Other problems with ventilation systems are less obvious, involving cross-talk, break-out or flanking via the duct.

Cross-talk becomes a possibility when a noise-sensitive space lies adjacent to a noisy area and where both areas are served by the same ventilation duct. If the separating wall has an insulation of approximately 40 dB(A) or above, then speech or noise may pass via the grilles and duct from one room to the other and reduce the effective insulation. The solution depends on the magnitude of the problem, but will vary between simply applying a lining between the grilles and creating separate branches to each room.

Break-out is normally associated only with rectangular ducting, although it may be extremely serious at the 'ring' frequency of large (> 1.0 m dia.) circular ducts.

In theory, there are four methods of reducing break-out, namely, stiffening the ductwork, increasing the mass of the duct wall, applying damping compounds or applying acoustic lagging. The only effective treatment under normal circumstances is the use of lagging, but space restrictions may prevent this method being applied.

One of the most complex theoretical problems involving ductwork has been the effect of flanking via the closed duct, that is, not via the grilles. In many buildings, the plantroom lies adjacent to a noise-sensitive space for which a heavy concrete wall has been designed to isolate the transmission of noise. When a duct is passed through such a wall, even although it is closed and en route to another room, a reduction in overall insulation may exceed 20 dB, with serious consequences. The solution is to apply acoustic lagging along the whole length of the duct within the room. A design procedure to predict the need for this is now available [10].

A similar flanking problem to duct-borne transmission is that via wall-mounted radiators which passes along the external facade by passing the insulation provided by the separating wall. Generally, the application of an acoustic barrier within the heater will suffice here, but care must be taken to ensure that adequate sealing is applied at the heater/wall junction.

An increasing problem involves the transmission of noise from externally situated plant such as cooling towers, auxiliary generators, etc., to adjacent dwellings. Very often, the problem here is the application of spherical attenuation guidelines, when hemispherical or quadrantal radiation actually exists. This, once again, is largely a fault of educational establishments which tend to instill in students' minds that sound radiates in accordance with the Inverse Square Law at 6 dB per doubling of distance, regardless of source or path characteristics.

4. DESIGN LIMITATIONS – INSTALLATION

Undoubtedly, the commonest reasons for excessive noise transmission result from incorrect installation, inadequate commissioning or, simply, downright bad workmanship.

With brick party walls, the insulation may be reduced by the practice of laying bricks with the frogs down, instead of up, thus giving rise to air voids and a reduction in the mass of the wall.

Partly filled joints in both brickwork and blockwork are commonly found in the part of the wall above the suspended ceiling. This area is rarely plastered and therefore has little resistance to air flow.

Inadequate sealing around service ducts and pipes where they pass through walls or floors can often be a cause of leakage.

Leakage often occurs at butt joints between walls and floors. Problems occur most frequently at the junction between the external wall and an internal partition, or at the wall head of a non-loadbearing partition. Cover plates and cornices give only limited insulation at these junctions and it is important to minimise the gap between elements, and to pack the space with mineral wool.

Air gaps are often found between door frames and the wall reveal or, alternatively, at the foot of the door. These air gaps can reduce the effective insulation of a door to less than 10 dB(A), instead of 20–30 dB(A), and thus give little privacy to the room occupant.

The sound insulation of demountable partitions is particularly affected by the accuracy of fixing the frame to the surrounding structure. Indeed, all frame partitions require to be well sealed at the junction between the frame and the structure, by either mastic or compressible non-porous rubber. A gap of one millimetre can reduce the insulation by up to 4 dB(A).

5. FUTURE REQUIREMENTS

The general solution to at least part of the problem of poor acoustic design in buildings lies in a better education in the subject at college and university level. Lecturers must acquaint themselves with the concepts of *practical* noise control, in addition to the theoretical laws upon which so much reliance is still placed.

Better instructions to installation engineers and craftsmen, and closer supervision of acoustically significant details, will reduce the incidence of excessive noise. If rigorous checking of all noise criteria levels before hand-over were to be introduced as standard procedure, then perhaps more attention to acoustic detail would follow automatically.

In more specific terms, equipment manufacturers should be required to produce both sound-power and vibration levels in octave band-width for every item of noisy plant.

Tighter specifications relating to workmanship and details should be laid down, describing the sound-transmission loss and absorption coefficient of materials, so as to reduce the possibility of misinterpretation of data by designers.

Finally, it should be said that industry will not adopt these procedures or produce the detailed specifications required unless pressure is brought to bear by professional organisations such as the RIBA or CIBS. We are not dealing here with problems at the frontiers of science, rather it is a case where most of the solutions are available. It is only a question of direction and enforcement that is required.

REFERENCES

1. *Environmental criteria for design*, Guide A1, CIBS (1978).
2. *Code of Practice for reducing the exposure of employed persons to noise*, HMSO (1972).
3. Mackenzie, C.M., and Mackenzie, R.K. 'The effects of noise upon clerical efficiency', *Proc. of the Conference on Building Acoustics*, Institute of Acoustics (August 1979).
4. *The influence of materials and construction on the acoustic climate in dwellings and its effects on residents' health*, Ministry of Building, Poland (1975).
5. BS 4196: 1967, *Guide to the selection of methods of measuring noise emitted by machinery*, British Standards Institution (1967).
6. Mackenzie, R.K., and Williamson, J.J. 'Sound insulation of lightweight concrete', *Build International*, 244–52 (1971).
7. Craik, R.J.M. 'An investigation in structureborne flanking transmission in buildings', PhD thesis, Heriot-Watt University (1980).
8. Sewell, E.C., and Utley, W.A. *The effect of resilient fillings on direct and flanking sound transmission with cavity masonry walls*, Building Research Establishment, Current Paper CP10/75.
9. Mackenzie, R.K., and Williamson, J.J. 'The cost effectiveness of double glazing against traffic noise', *Build International*, 127–47 (1976).
10. Craik, R.J.M., and Mackenzie, R.K. 'The transmission of sound by ventilating ducts – a design guide', *J. Applied Acoustics*, **14**, 1–5 (1981).

9 Case study – some noise control problems encountered in the design of a local-authority housing development

Ron Simpson

OBJECTIVES

The subject of this case study is Grafton 3a, a new housing development commissioned by the London Borough of Camden in 1973, comprising 122 dwellings and covering an area of 1.6 hectares in Kentish Town. In 1972, a year before work began on this project, the DOE published Design Bulletin 26 'New housing and road traffic noise, a design guide for architects', and this site undoubtedly represents an extreme example of the problem envisaged in the bulletin. In this chapter the measures are described that were adopted to reduce the noise within the housing development to acceptable levels, within the stringent cost limits governed by the Housing Cost Yardstick.

Figure 9.1 Site relation to external noise sources

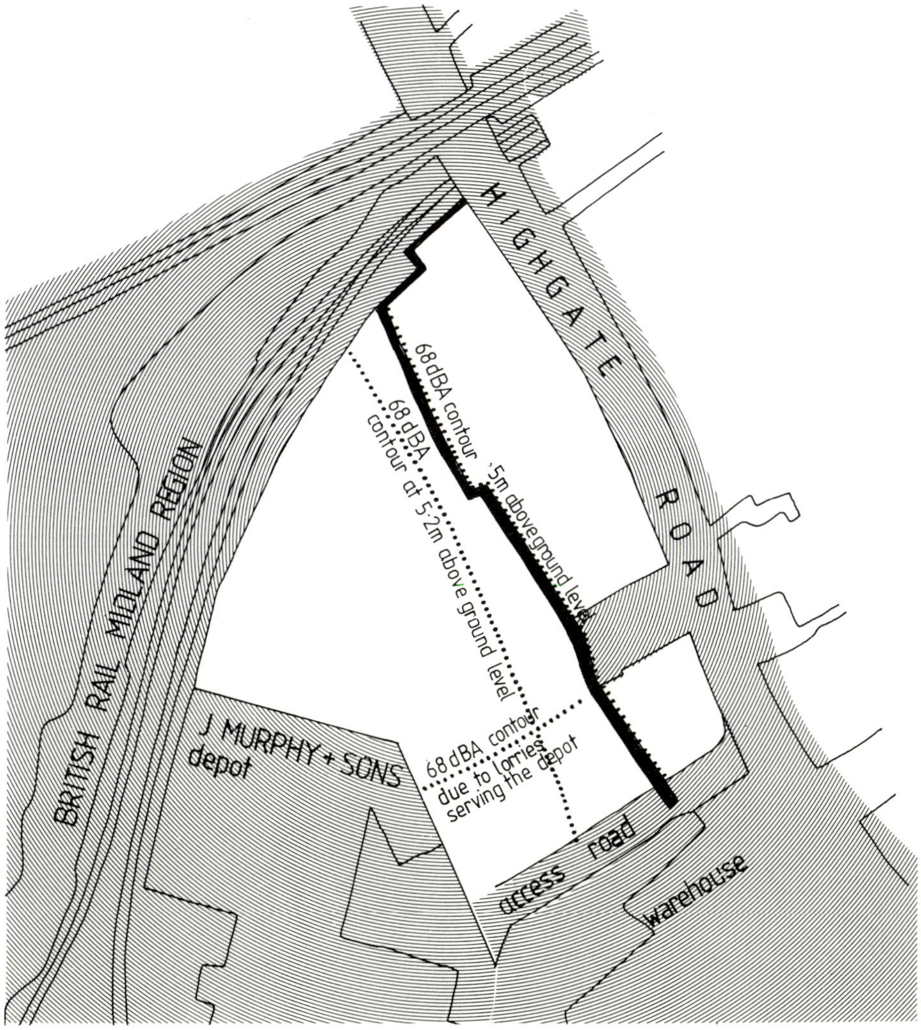

Formerly a locomotive depot, the site occupies a deep hollow to the west of Highgate Road and is bounded by a large contractor's depot to the west, two railway lines, one high- and the other low-level, to the north, and a large warehouse and a road serving the depot to the south (see Figure 9.1).

We were required by Camden to prepare a noise survey and for this, and subsequent advice, Sandy Brown Associates were appointed consultants. We are particularly indebted to Dr Barron who was largely responsible for the detailed work on the acoustic design.

The noise survey was made in January 1974, using a microphone at a height of one meter in the open. The dB(A) level was continuously recorded on a Band K 2305 level recorder for the whole period and supplementary tape recordings for frequency analysis were made of significant noise sources. Figure 9.2 represents the result of the survey and the following principal sources of noise were observed.

Figure 9.2 Result of noise survey
A Peak level of more than 2 seconds duration due to lorries serving depot
B Mean level from traffic in Highgate Road 2 m from kerb
C Background spectrum at 24-hour measuring position
D Peak levels of wheel squeal due to railway

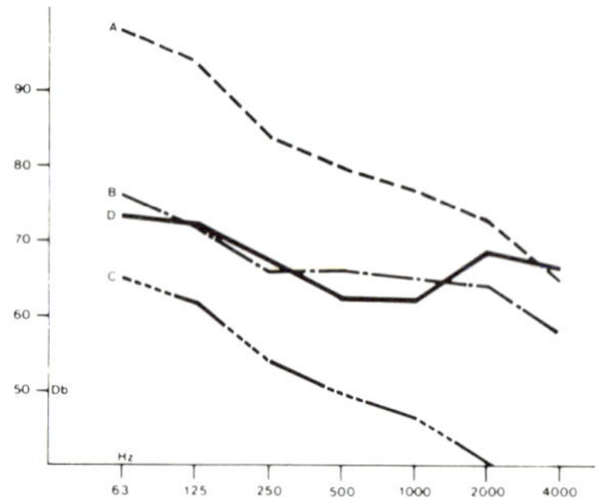

1. RAILWAY

Trains passed along the lines during the whole of the 24-hour period and the majority were goods trains travelling at slow speed and, as the line was on a curve, there was frequent wheel squeal. The railway noise is only likely to cause annoyance at night-time, during which period a frequency of 4–5 trains per hour may occur. At the 24-hour measuring position, peak levels above 70 dB(A) were measured during the night.

2. ROAD TRAFFIC NOISE

Traffic counts indicated a mean daytime rate of 110 vehicles per hour, rising to 1350 during the rush hour. The L_{10} dB(A) level at a position 2 m from the curb during daytime was 75 dB(A); the spectrum is given in Figure 9.2. Night levels can be estimated at 10 dB(A) lower.

3. J. MURPHY DEPOT

The background noise spectrum adjacent to the depot entrance indicated an L_{10} level of 62 dB(A). However, the noise of trucks serving the depot created a more serious nuisance. Between 50 and 60 trucks leave the depot between 06.30 and 08.00 hours each morning, which means an L_{10} level of about 76 dB(A) during this period adjacent to the access road. The trucks return between 16.00 and 18.00 hours. Sandy Brown's recommendations were as follows:

Using the Wilson report as a basis, the locations at which the L_{10} level exceeds the maximum of 50 dB(A) during the day and 35 dB(A) during the night would require special consideration.

In the central area, the railway noise exceeded the limit by as much as 12 dB(A) and it was recommended that a row of buildings should run parallel to the railway to act as a barrier to the railway noise.

The road traffic noise exceeded the limit by as much as 25 dB(A) at night and it was recommended that any housing close to Highgate Road demanded special acoustic treatment to habitable rooms.

Noise levels produced by trucks serving the depot in the morning posed a serious problem and considerable shielding and separation were recommended.

As the housing development was for a local authority, the cost limit was strictly governed by the Housing Cost Yardstick as determined by the Department of the Environment. It should be pointed out that the cost of measures necessary to control noise is acknowledged by the DOE and a special grant for this purpose is assessed in repect of all parts of the site coming within a 68 dB(A) noise contour.

This figure appears to be related to 'Building and buildings' – The Noise insulation Regulations 1973 Statutory Instrument No. 1363, which refers to noise from highways. The relevant 68 dB(A) level is the L_{10} level for the 18-hour period 06.00 hours to 24.00 hours, while the problem on this site is likely to be governed by night-time noise levels, if Wilson Committee requirements are to be fulfilled.

A further complication was caused by the dual levels of the site. A substantial area is outside the 68 dB(A) contour at ground level, because of the shielding effect of the two-storey-high retaining wall, but exceeds it at heights above the upper Highgate Road level. A contour in respect of the road serving the traffic from the depot could not be included, since it only applied during four hours of the day and the 18-hour L_{10} criterion did not apply and was therefore not recognised as a nuisance for special grant purposes. A 68 dB(A) contour map was prepared, see Figure 9.1, estimated for the period 06.00–24.00 hours. The contours in respect of road noise were based on measurements, while the interaction zone where there were contributions of noise from both Highgate Road and the railway were calculated according to paragraph 39 of the Design Bulletin 26 as recommended in the statutory instrument.

Thus, in agreeing the Housing Cost Yardstick with the DOE, the allowance in respect of the allowable noise within the 68 dB(A) contour amounted to £97 000, a sum representing £790 per dwelling, compared with the remaining yardstick for the entire development of £1 291 000. The density adopted by the Camden planners for this site was 250 persons per hectare and our quantity surveyor's advice on solutions viable within the cost yardstick pointed towards a relatively low-rise solution not exceeding four storeys and generally of loadbearing brick construction.

We welcomed this approach, since we were anxious to create an environment suitable for families and children, with as many front doors as possible at ground level, and with full use of open space and private gardens. Thus, it was in this context that the noise-control measures were studied.

The solution adopted exploited the pronounced change of levels and, by placing a four-storey maisonette block close to the retaining wall dividing the site, gave rise to wide-fronted dwellings with all habitable rooms facing in one direction towards the sun, while stairs, bathrooms, and stores occupied the dead north-east side, with entrances at the upper street level, see Figure 9.3.

This dwelling type, with its relatively blank entrance-front facing the noise source, and rooms facing the central protected area, created the ideal form to act as a noise shield and was used in a modified form along the railway boundary. In the detail design, a mono-pitch roof form was adopted to maximise the protective wall facing the noise source, see Figure 9.4.

On Highgate Road, the brief called for a row of shops and we chose to plan the required flats over them. Because of limited street frontage and the requirement for a minimum of two hours' sunshine to living rooms, a narrow-fronted plan was designed, with bedrooms facing north-east towards Highgate Road. The windows to the bedrooms in this situation required a performance of 35 dB(A) to satisfy the Wilson Committee recommendations. Tenders were invited for windows, based on a performance specification, and the window selected was a timber tilt-and-turn double casement, manufactured by Messrs Ripper.

Figure 9.3 Birds' eye view of development

Figure 9.4 Section through screen block

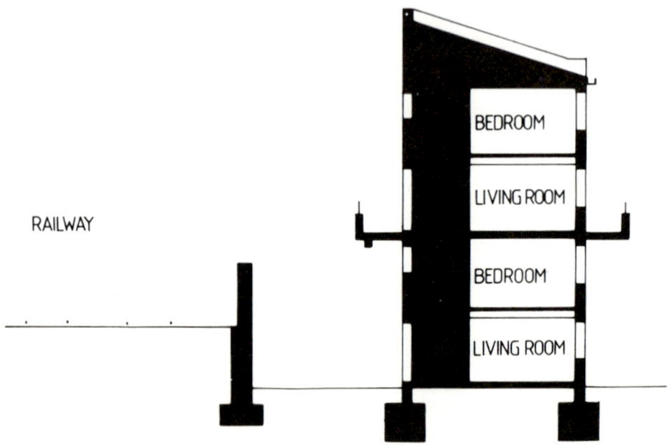

Because of the need to conserve energy, all the windows are of minimum area and are double-casement for thermal insulation. The meeting faces have integral neoprene gaskets and the tilt-and-turn form allowed total contact on all four sides. Softwood was used, stained rather than painted, to preserve the neoprene contact faces, and for the bedroom windows facing Highgate Road an extra deep version was designed, with a 100 mm gap glazed internally with 6 mm and externally with 4 mm glass, see Figure 9.5. For the rest, however, windows of identical construction, but with a 40 mm air gap, are used.

Messrs Ripper were required to carry out a test on a completed 100 mm window and this was done by Sound Research Laboratories Limited at Sudbury, Suffolk. Measure-

Figure 9.5 Detail of special window

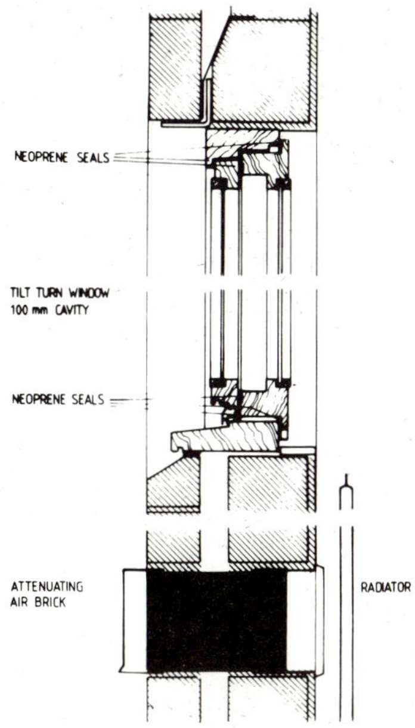

Figure 9.6 Result of acoustic performance tests on special window

- **A** Laboratory test of basic window with absorbent lining
- **B** Laboratory test with additional seals to all four edges and with lining removed
- **C** Sound-attenuating air brick, manufacturer's figures
- **D** Camden Site survey with air brick sealed
- **E** Camden Site survey with air brick open

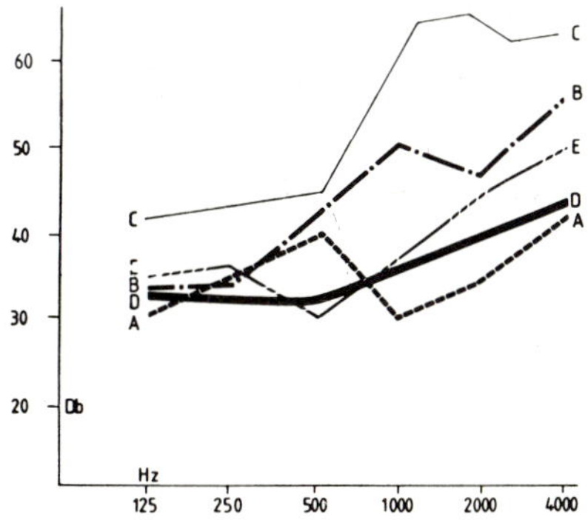

ments were conducted on the basic window in accordance with BS 2750: 1956, and a number of modifications during the testing were made and the results monitored, see Figure 9.6.

The basic window, fitted with 12 mm fibreboard lining to the reveal, gave an average result of 33.6 DN. Adding neoprene seals to head and sides of frame between the two sashes gave a result of 34.4 DN, and adding a seal to the aluminium sill section gave a result of 38 DN. At this point it was decided to see the effect of omitting the fibreboard lining and this gave a result of 33.7 DN. Further neoprene seals were added to the room face of the window and this gave 40.4 DN. It was in this final form that the window was eventually used.

With the window in a closed position, a secondary form of ventilation is necessary, and under the London Building Act it is required to provide at least three square inches of permanent ventilation in all habitable rooms. Electrical fan ventilator units were investigated but at that time there was no model available which satisfied the GLC requirement for permanent ventilation.

We discovered, through enquiries with the Gas Council, that Oldham Sheet Metal Co were manufacturing special sound-attenuating air bricks for admitting air for gas fires in a housing scheme adjacent to a trunk road in Stockport. The air brick is made of lead, with perforated tubular air intake surrounded with a metal wool. The outer face is protected by a stainless steel baffle screen. Noise test results produced by the manufacturers are shown in Figure 9.6.

The air bricks were installed below the window, immediately behind the hot-water radiator, with the result that the incoming air was warmed and probably deterred the tenant from blocking the grille.

In the central part of the site are located large two-storey family houses and, in dealing with the industrial and depot traffic noise, there seemed to be only one course of action, because of the orientation: to erect a substantial barrier at the boundary and locate the dwellings as far away from this as possible. A 5 m solid brick wall, 400 mm thick, was located on the boundary with the depot, with a similar 3 m solid brick wall along the boundary with the depot road. The road rises steeply, which probably helps the screening effect of the wall by increasing the angle of deflection of the noise path relative to the dwelling windows. In addition, the nearest dwelling is about 20 m from the centre of the road and some reduction can be expected due to distance. Finally, because of the combination of solid barrier and reduction due to distance, it was felt that the special 100 mm windows were not justified for these dwellings, and the standard type has been used.

On 26th October 1979, a survey of the noise condition in the bedrooms facing Highgate Road was carried out by the Environmental Health Department of the Borough of Camden. Measurements were taken simultaneously inside the bedroom and externally, and these are represented in Figure 9.6. It was found that the attenuation provided by the window was virtually the same whether the acoustic air brick was sealed off or left open, indicating that its performance was satisfactory for this use. The attenuation provided by the window was between 34 and 35.6 dB on average, between 125 and 4000 Hz, which achieves the recommendations of the Wilson report.

At the time of writing, the development is not occupied and it is expected that tenants will move in in January 1980. It is proposed to interview tenants after several months' occupation, to establish the effectiveness of the sound-control measures in a living environment.

Meanwhile, with the building work complete and landscaping well under way, the general impression within the housing is that much of the external noise has been reduced to a tolerable level and within the dwellings it has been practically eliminated.

10 Sound in large rooms

Brian Day

1. CONCEPTUAL MODELS FOR ACOUSTICS

Whenever we wish to describe a system mathematically, we need to start with some conceptual model of the process that we are describing. This is true for room acoustics: from the very earliest times the design of theatres and other spaces has been determined by some picture in the designer's mind of the way in which sound behaves. This review of theories of sound in rooms presents the models which can give guidance to designers and their advisers at all stages.

The models fall into three main classes: the wave model, in which the motion of the air in a room is described in terms of the superposition of pressure waves travelling through the air; the geometric model, in which sound is assumed to travel around a room like rays of light, being reflected at any surfaces that it encounters; and the 'energy' model, in which sound is treated simply as a flux of energy which is stored within the confines of the room. For any one problem, the description of sound in rooms may make use of any or all these models and there are benefits or disadvantages to be considered in choosing a particular approach.

One should perhaps consider briefly the need for appropriate 'models' of any kind. It might be argued that the building scientist should attempt to provide as near a description of the real world at as fundamental a level, and as explicitly, as possible. One can conceive of a complete mathematical formulation of all the boundaries of a room, including their acoustic properties, and of the sound waves that might be propagated in the medium filling the room. However, even if one were to write down the differential equations and the boundary conditions in a detailed enough way, the prospect of achieving a full solution of these equations would be remote, in spite of the potential use of large computers. Furthermore, the ability of the occupants of a room to perceive its acoustic properties is limited by the nature of aural perception and not least by the fact that any auditor can only be in one place at a time. Thus, a full description is neither necessary nor possible. But the use of the appropriate model may convey all the significant information about the room in a way that makes it possible for the designers to design for people. If building designers were punters at a race meeting, they would not be dissuaded from a study of the form book by the fact that a complete description of the molecular construction of every horse was not available. The fact that we can conceive of a complete description of the acoustic properties of a room, even though such a description may not be achieved, should not prevent us from searching for an appropriate 'summary' model which will help designers to design better buildings.

2. THE WAVE MODEL

The earliest statement of the wave model of room acoustics occurs in a famous passage in Book Five of Vitruvius' Ten Books on Architecture. In Chapter Six he gives us the following explanation:

Voice is a flowing breath of air perceptible to the hearing by contact. It moves in an endless number of circular rounds like the unnumerably increasing circular waves which appear when a stone is thrown into smooth water and which keep on spreading indefinitely from the centre unless interrupted by narrow limits or by some obstruc-

tion which prevents such waves from reaching their end in due formation. When they are interrupted by obstructions the first waves flowing back break up the formation of those which follow.

In the same manner the voice executes its movement in concentric circles; but while in the case of water the circles move horizontally on a plane surface the voice not only proceeds horizontally but also ascends vertically by regular stages . . .

Vitruvius then proceeds in a completely erroneous analogy to describe how the proportion of theatres should be determined by the rules of Greek harmony, but this should not detract from our astonishment at the perceptiveness of his insight. There is no way in which the wave nature of sound could have been observed physically except through a leap of intuition which related the diffraction of water waves around obstructions with the way in which sound could be heard even when direct sight was blocked.

In a later chapter Vitruvius goes on to categorise the different kinds of reflection that can occur in a theatre (the 'dissonant', the 'circumsonant', the 'resonant' and the 'consonant'). Although Vitruvius' model of the cause of these phenomena is in error, it is quite possible to identify the phenomenon to which he refers. His model seems to involve interaction between waves that have travelled out from the person speaking on the stage of his theatre and being reflected from various surfaces and falling back and 'suppressing' succeeding waves. Because his description deals with design for speech (or perhaps singing) in particular, he is concerned with transient sounds rather than the steady-state conditions which determine noise problems. What he calls 'resonant' reflections we would now refer to as 'distinct echoes', while his 'consonant' reflections probably correspond to those early reflections which reinforce the direct sound. 'Circumsonant' reflections seem to correspond to what we would now refer to as 'reverberation', while the 'dissonant' is the kind of reflection which Vitruvius thought was responsible for suppressing the subsequent sound waves.

It is worth taking time to consider Vitruvius' views of acoustics, because they give us an opportunity to learn to avoid a rather common failing in design theory. Although Vitruvius gives us a plausible classification of the kinds of reflection that might be obtained in a theatre and tells us which reflections are beneficial and which detrimental to the acoustics of the theatre, he gives us no guidance as to how these reflections are to be obtained or prevented. He merely tells us that:

. . . hence if there has been careful attention in the selection of the site the effect of the voice will through this precaution be perfectly suited to the purposes of a theatre.

This is not enough: the building scientist should not just ask himself, of a new model, 'Is it true?' and 'Is it verifiable?' He should also ask, 'Will it produce useful advice for the designer?'

Further progress in the application of wave theory to our understanding of the acoustics of rooms had to wait until the middle of the nineteenth century, when Lord Rayleigh published the theory of standing waves in a three-dimensional enclosure. The wave equation, which relates sound pressure p to position x and time t, takes the form (in one dimension):

$$\frac{d^2p}{dx^2} = \frac{1}{c^2}\frac{d^2p}{dt^2}$$

In three dimensions this becomes:

$$\frac{d^2p}{dx^2} + \frac{d^2p}{dy^2} + \frac{d^2p}{dz^2} = \frac{1}{c^2}\frac{d^2p}{dt^2}$$

where x, y and z are distances along the three axes at right angles to each other.

If we impose boundary conditions on the solution of this equation, which incorporate the dimensions of the room and the nature of the boundary walls, then only certain distributions of pressure (p) are presumable, corresponding to fitting in certain standing waves within the distances available between the walls. For a rectangular room with hard walls the solutions are of the form:

$$p_a = \sum_p \sum_q \sum_r C_{pqr} \cos p\left(\frac{\pi x}{a}\right) \cos q\left(\frac{\pi y}{b}\right) \cos r\left(\frac{\pi z}{c}\right)$$

where the dimensions of the room are a, b, c, and p, q, r can take any integer value. C_{pqr} is the pressure amplitude of a particular 'mode'. The frequencies which correspond to these modes are given by the equation:

$$f = \frac{c}{2}\sqrt{\left(\frac{p}{a}\right)^2 + \left(\frac{q}{b}\right)^2 + \left(\frac{r}{c}\right)^2}$$

The degree to which the room can sustain sound at other frequencies will depend on the 'sharpness' of these natural modes. If the walls are extremely hard and reflective, then only frequencies which match the modes precisely will be strongly sustained within the room; other frequencies will not be radiated strongly by sources, since the room will present a mismatch in impedance at the source. As anyone who has ever sung in a bathroom will know, certain frequencies can be excited strongly in a small room. This occurs because the resonant modes present a powerful impedance at the source of the sound, enabling it to transfer large amounts of energy into the room.

The values of p, q, and r in these equations specify the modes of vibration. For example, if $p = 1$, $q = 0$ and $r = 0$, this describes a standing wave with a single 'loop' in a direction parallel to the x axis, that is to say, the direction in which a is measured. Higher order modes would have higher values of p and correspondingly greater numbers of 'loops'. Mixed modes are permitted in which more than one of p, q or r is non-zero; for example, the mode with p, q and r all equal to one describes a standing wave which runs along the body diagonal of the rectangular room.

If we now consider the frequencies at which these natural modes of vibration can occur, we find that it is only in very small rooms that the frequencies are sufficiently widely spaced to have an observable effect on the acoustic properties of the room. Generally speaking, the resonances at normal frequencies and in normal-sized rooms are so closely spaced that their response curves overlap and discrete resonances are not heard. However, if we carry through the calculation for a room of a few metres square, we find that the natural frequencies are widely enough spaced to give discrete peaks in the overall response curve. In this region of behaviour, the noise level produced by a source operating within the room will be affected by the proximity of the frequency emitted by the source to these natural modes, and to the degree to which they are damped by the acoustic properties of the boundaries of the room. In the derivation of the equation above for the pressure distribution it has been assumed that the walls are 'infinitely hard'. If realistic boundary conditions are applied, incorporating the acoustic impedance of the surfaces, it is found that the acoustic impedance presented at a source varies less with frequency and location in the room, consequently smearing out the response curve predicted for infinitely hard walls, and limiting the sound-pressure amplitude generated by a source to less than infinite values.

A further problem with the application of the wave theory to rooms (apart from the fact that extremely large numbers of resonances are predicted for normal-sized rooms) is that the fundamental assumption of rectangularity of the room is not often fulfilled. As soon as one moves away from plane surfaces intersecting at right angles, the mathematics of the solution of the wave equation become far too complicated to deal with. The precise application of the equation is therefore not possible, although an indication of the resonant behaviour can be obtained by approximation. Some impression of the complexity of the solution that would be called for if a complete analytical approach were adopted can be gained from the fact that a room with a volume of 3000 cubic metres would have no less than 1000 separate modes of vibration in the frequency range from 1000 Hz to 1001 Hz. Although there has been some success in recent years in extracting simplified descriptions from the full, rigorous wave theoretical treatment of acoustics in rooms (notably by Morse and others), this has largely been useful only in so far as it demonstrates that the wave theoretical approach leads to the same results as the more direct reverberation approach to be discussed later. Anyone wishing to pursue this topic is referred to reference [1] (Morse) or reference [2] (Kuttruff). The most likely area in which the theory might be of use to noise control practitioners would be in the circumstances where a sound source has a particular frequency and is enclosed in a small room; it is probably worth considering the possibility of the coupling of the single-frequency source with a natural mode in the room.

3. GEOMETRICAL MODEL

Probably the simplest conceptual model for describing the behaviour of sound in large rooms is that which treats the sound as if it consisted of rays of sound energy obeying the same geometric laws of reflection as does light. This in a sense ignores the wave nature of sound in that diffraction effects are completely neglected. Although the concept itself is simple, the techniques which can be based on it increase in complexity year by year. One of the earliest examples of the use of this geometric approximation gave rise to a design form used extensively in the Chicago Auditorium, by Sullivan. The concept was that of the Isacoustic curve, devised by the engineer John Scott Russell. This proposed that good acoustic conditions would be obtained if the angle of incidence of sound rays from a speaker on to the audience plane were maintained constant at different distances from the speaker. There is a phenomenon (now known as the Bekesy effect) by which sound travelling close over the heads of an audience is rapidly attenuated even though there is no interposition of solid objects. The success of Greek and Roman auditoria was due in part to the avoidance of this effect with steeply raised seating areas. Scott Russell's idea was to maintain a constant angle of incidence for the sound rays and using this to arrive at curved sections for the auditorium floor, and for any balconies, which were determined exclusively by this constraint. The resulting so-called 'Isacoustic curve' was used in the Chicago Auditorium not only to generate the shape of the floor but also, less sensibly, to determine the curvature of the ceiling arches. These, in fact, are entirely applied, and have no structural function, consisting of lightweight lath and plaster construction. There is no reason to suppose that the acoustics of the hall were improved by the use of this particular curve in the surfaces other than the one on which the audience sat.

A more flexible application of the geometric acoustic theory to floor design is given by Ingerslev [3]. Rather than fix the angle of incidence of sound rays on to the audience and thereby determine a curve for the floor, he simply sets the degree of attack at a minimum value. This generates a lower limit for the step size between rows, rather than determining an exact curve.

A similar technique can be applied to the design of the shape of auditorium ceiling in section and side walls in plan. There is an inevitable weakening of direct sound as it travels towards the back of the auditorium, due to the inverse square law. This must be compensated by the provision of reflected sound coming from the ceiling or the side walls. It is possible to determine the shape which these elements of the room should take, in order to improve the acoustics in the rearmost seats of the hall, by simple geometric ray tracing. This principle was applied by Le Corbusier in his project for a debating chamber at the Salle des Nations in Geneva and carried to fruition in the Salle Playel in Paris, the first of the 'directed sound' concert halls. In this country, the Liverpool Philarmonic Hall is an example of an auditorium in which the shape of the ceiling has been determined so as to direct sound towards the rearmost seats, using the principles of geometric acoustics.

Another example of the way in which geometric acoustics can lead to simple design rules is in the study of the focussing of sound by concave surfaces. The geometric acoustic analogy with optics leads us to expect a concentration of sound energy at the focal point of a curved surface and indeed this is often observed in practice. The phenomenon known as the 'whispering gallery' effect is also explained by this simple model. It can also predict the repeated reflection from parallel facing walls which gives rise to flutter echo and the long detailed reflection which results from a tortuous path amongst a number of surfaces to give rise to a discrete echo in large rooms. Ray-tracing techniques thus form an important part of the designer's tool kit, if only because they are so easy to apply.

With an extension to cope with diffraction effects, the geometrical approximation can also be used to calculate the effects of barriers and overhead reflections on received noise. For example, if we consider a noise source within a factory which is to be provided with a screen, by tracing the rays from the source which might be reflected from the ceiling or diffracted over the top of the screen we can make simple calculations to predict the resulting sound energy which arrives at any point on the factory floor. Although the calculation of the proportion of energy diffracted should depend upon the wave length of the sound (and thus on a wave model), it is possible to use a rule-of-thumb computation technique to arrive at a figure for the energy travelling in this way.

By the use of the concept of absorption coefficient, it is possible to extend the model further, to make calculations of the sound energy reaching a point by reflection from nearby surfaces, traced by the geometrical approximation. Absorption coefficient is defined by the equation:

$$a = \frac{\text{energy not reflected}}{\text{energy incident}}.$$

The location of an image can be obtained by graphical construction. If the inverse square law is then applied from the image to the point of observation and the strength of the image reduced by the application of absorption coefficient to its power, then the pattern of energy arriving by a number of routes can be calculated. This geometric model of course neglects the effect of interference, which would only be treated from a wave theory viewpoint. Nonetheless, useful results can be obtained by the application of this technique, especially now that use is being made of computer programs to follow the multiple interreflection of sound rays between the surfaces of quite complex interiors. One program has been developed to trace sound rays through a number of reflections in a concert hall, and this can be used at the early stages of design to compute the pattern of reflections that arrive at any seat in the auditorium [6].

The geometrical model of room acoustics can be realised physically by the use of the optical analogue technique. The technique, developed first by Phillips Laboratories [4,5], involves the construction of a model of an interior in which the surfaces are made with optical reflectance equal to the acoustical reflectance (l-absorptance) of the surfaces in the full-size prototype. A light source is located in a position corresponding to that occupied by the acoustic source in the real building. By using a light detector to measure the level of the illumination at points on the 'floor' of the model auditorium, an investigation can be made of the distribution of sound to be expected in the prototype. Investigations can also now be made by means of a low-power laser beam. Because they are so tightly collimated, they provide a three-dimensional representation of the geometric model and enable the physical realisation of the ray-tracing technique.

It should be apparent that there are two essentially different ways in which the geometric model can be applied. In the first application, consideration is given to the total sound power developed by a continuous source, as, for example, in a factory. It is possible to use the geometric model to calculate the location of the images of a single source, and to superimpose the power received from the source and its images, using the inverse square law to calculate the resulting total sound-pressure levels. Any barrier in the way of a direct or reflected path can be treated by using an empirical diffraction formula. Absorption at a surface is taken into account by reducing the intensity of the image in proportion to the absorption coefficient of the surface through which it is 'seen'. Multiple (higher order) images can also be treated by considering the original image to be reflected again in another surface of the room. An example of this technique, applied to landscaped office studies, appears in [7].

The second way in which the geometric model can be applied is to the prediction of transient acoustic behaviour in rooms. The main application is in considering the reinforcement of direct speech by early reflections. The Dutch acoustician Haas demonstrated (using electroacoustic techniques) that any repetition of an original sound picked up within 50 milliseconds of the arrival of the direct sound is not detected by the hearing process as a separate, distinct repetition. Instead, the energy is integrated by the brain so as to raise the apparent level of the direct sound. Sound arriving after 50 milliseconds is then either distracting or confusing sound, depending on the degree to which it is delayed. Using this as a basis, it is possible to test particular reflections, traced by the geometric method, to see if they will contribute or detract from the clarity of speech in any part of an auditorium. The inverse square law can be used to compute the relative intensity of the reflections, their delay can be estimated from the path length and hence their significance in providing clarity of speech in an auditorium may be determined.

4. THE STATISTICAL (ENERGY) MODEL

The foregoing models treat the sound either as a wave motion setting up resonances within the room or as 'particles' of sound travelling in straight lines between reflections. The statistical, or energy, model treats the sound field as the result of the emission of sound energy from the source into the room, which is regarded as a single store filled uniformly with sound energy which 'leaks' away at the walls of the room. This model, first introduced by Wallace Sabine in 1900, takes the view that a sound source operating within a room fills up the room with sound energy. After the source is cut off, the sound energy leaks away at the wall and this can be heard as a prolongation of the source. Sabine's original experiments used organ pipes with air pumps and a stop-watch to determine experimentally the time taken for the sound to die away. First using one hall and varying the amount of absorption and subsequently carrying out the experiments in other halls, he established that what he called the reverberation time (the time for the intensity level to fall by 60 decibels) was related to the volume and the absorption on the surface of the room by the equation:

$$T = \frac{0.16 \times \text{volume of room}}{\text{total sound absorbing power}}.$$

This experimentally determined equation was subsequently derived theoretically by Jaeger, who treated the sound in the room as a 'gas' of phonons which moved at random through the room and were absorbed whenever they struck a surface with an absorption coefficient. Jaeger applied the theory of absorption of gas molecules from a vessel being evacuated with a vacuum pump, in order to show that Sabine's result could be derived theoretically. There is, unfortunately, an apparent anomaly in that if the absorption coefficient in Sabine's equation is to be equal to one (that is to say, if all the walls of the room are covered with a perfectly absorbing material), the reverberation time predicted does not go to zero. This is not so unsatisfactory as it seems at first sight, since it represents the average time for sound energy to travel from the source to the wall, where it is immediately absorbed. However, Norris and Eyring independently derived expressions which did not suffer from the defect. Eyring assumed the sound field in a room to be the sum of that due to the source and all the images that it forms in the walls by multiple reflection. When the source is switched on, the model assumes that all the images switch on simultaneously and the development of the sound field in the room then reflects the arrival of the sound from these image sources from increasingly greater and greater distances. The strength of each of the image sources is weakened by the fact that they are generated by reflection from the imperfectly reflecting walls of the room. This results in an infinite series:

$$E = \frac{W}{4\pi r_0^2} + \frac{\bar{a}\,W}{4\pi r_1^2}\,n_1 + \frac{\bar{a}^2\,W}{4\pi r_2^2}\,n_2 + \ldots\ldots\ldots \frac{\bar{a}^N\,W}{4\pi r_N^2}\,n_N$$

where W = source power
n_N = number of Nth order reflections
r_N = average distance to Nth order reflections
$\bar{a}$ = average absorption coefficient.

This leads to the equation:

$$T_{60} = \frac{0.16 \times V}{-\,S_{\text{total}} \times \log_e (1 - \bar{a})}$$

The Norris approach assumes that a ray of sound travels around the room being reduced by $\bar{a}$ on each reflection. (The average distance traversed by a 'sound particle' between reflections is called the 'mean free path'.) Since it suffers a reflection on average each time it traverses a 'mean free path', the time interval between reflections equals the mean free path divided by C, where C is the velocity of sound. The energy of the ray is reduced to:

$$E = E_0\,(1 - \bar{a})^{\frac{\text{m.f.p.}}{C}\,t}$$

It can be shown that the mean free path is given by:

$$\text{m.f.p.} = \frac{4V}{S}$$

so that the energy of the 'sound particle' is thus given by:

$$E = E_0 \exp\left[\frac{Sc}{4V} t \log_e (1 - \bar{a})\right]$$

which eventually leads to the same expression for the reverberation time.

A further approach adopted by Millington for multiple surfaces gives an expression:

$$E = E_0 (1 - a_1)^{N_1} (1 - a_2)^{N_2} (1 - a_3)^{N_3} \ldots$$

where the N's in the expression are the numbers of reflections from surfaces S1, S2, etc. If we assume that the number of reflections is proportional to the proportion of the total area represented by a particular area, then this expression leads to:

$$T_{60} = \frac{0.16\ V}{-\ \Sigma S_N\ [\log_e (1 - a_N)]}$$

Generally speaking, these expressions will not give answers widely different from each other. However, for rooms in which there is a very high absorption coefficient on average, or rooms that depart widely from having simple proportions it may be necessary to proceed to more complex formulae; these are discussed very fully by Kuttruff in [4].

The statistical model, for the treatment of steady-state density, where a source operates continuously, was also derived by Jaeger and has subsequently been derived by others. The simplest expression for the way in which the steady-state sound-energy density can be calculated emerges directly from these equations: it is necessary only to substitute a large value for t. This gives rise to the simple expression:

$$E = \frac{4W}{R}$$

where E is the reverberant, steady-state energy density, W is the sound power of the source (in watts) and R is the so-called room constant given by:

$$R = \frac{\Sigma S\bar{a}}{(1 - \bar{a})}$$

5. PHYSICAL MODELS

In addition to these conceptual and mathematical models it is possible to construct both analogue and physical models for investigating the behaviour of sound in rooms. The optical model has already been mentioned, in which light rays are taken as an analogy for sound rays. Use has been made of electrical, resistance/capacitance analogues to study the growth and decay of sound fields in coupled rooms: the model is constructed from capacitors to represent energy storage (on the principle of the statistical model) in a room, with resistance representing the degree of acoustic connection between rooms. The response of the system to varying voltage applied to one point in the resulting network is then regarded as an analogue for the behaviour of the acoustic system in response to a varying sound input in one room.

Physical modelling of room acoustic behaviour has been established for several years and is now being applied regularly to the design of auditoria and even to factory spaces. The principle of the method lies in constructing a scale model (usually about 1/10 scale) of the proposed enclosure, finishing its internal surfaces with materials that have acoustic absorption properties matching those of the prototype but at scale frequencies, and then testing the model by conventional techniques, but at scale frequencies. Thus, in a tenth-scale model one would test at ten times normal audio frequencies. Much smaller models have been used for studying the propagation of traffic noise amongst buildings, but the major problem with this technique lies in the time taken to construct the models and the resulting mismatch with the conventional design process.

Simulation modelling with computer is my final example, and this still remains a potential rather than a realised technique. The computer-based acoustic model would consist of a set of instructions contained in a computer program which would represent the pattern of reflections and reverberation in a room. The sound wave spoken into a microphone would be digitised at very high speed and the resulting stream of data fed to the computer, which would then add the programmed reflections and reverberation before the resulting data stream was reconstructed into an acoustic waveform. This could then be played back, in real time, to the investigator. Unfortunately, the demands for speed and storage that this process would make are still excessive for existing computers, but it may be that it will eventually become possible to simulate and 'listen to' proposed buildings in this way.

6. CONCLUSION

An attempt has been made in this chapter to summarise the main models that can be applied in the description of the acoustics of rooms. Although we should not lose sight of the fact that some of the models are 'cruder' than others, the choice of model will often depend as much on the time and resources available for investigation as on the degree of precision required. A rough answer at the early stages of the design process may be very much more valuable than a more precise evaluation at a stage when it is too late for any amendments to be made.

In addition, the choice of model will also depend on the nature of the phenomenon and the space being investigated. It would not be appropriate to apply reverberation theory to an open-air theatre, for example. Room acoustic phenomena are much more complex than we perceive, and we do well to adjust our conceptual processes to match our needs.

REFERENCES

1. Morse, P.M. *Vibration and sound*, McGraw Hill (1948).
2. Kuttruff, H. *Room acoustics*, Applied Science Publishers (1973).
3. Ingersley, F. *Acoustics in modern building practice*, Architectural Press (1952).
4. Vermerlen, R., *et al. Phillips Technical Review*, **1**, 46–52 (1936).
5. Kuttruff, H. *Acustica*, **8**, 330–6 (1958).
6. Krokstad, A., *et al. J. Sound Vib.*, **8**, 118–25 (1968).
7. Day, B. *Nederlands Akoestisch Genootschap*, (23) 46–55 (1972).

11 Noise in factories

Andrew Middleton

SYNOPSIS

This chapter discusses noise sources in factories and how they should be described; effects of layout of workshops, plant-rooms, etc; noise control techniques available to the building designer; interaction between factory noise and operations and noise radiated into the neighbourhood.

1. INTRODUCTION

It is now common knowledge that we have a 'Code of Practice for reducing the exposure of employed persons to noise' [1], which recommends that employees should not be exposed to noise levels exceeding 90 dB(A) for 8 hours, or an equivalent continuous noise level of 90 dB(A) if the noise is not steady. While machine manufacturers have a large contribution to make towards the reduction of noise in factories, building designers can also make a useful contribution. If they fail to appreciate the problems of noise in factories, they can easily find themselves responsible for a disaster.

2. NOISE SOURCES IN FACTORIES

Obvious sources of noise are:

Machines
Ventilation or extraction fans.

Perhaps less obvious are:

Handworking operations such as hammering
Hand-held power tools
Materials handling
'Music while you work'.

In designing the acoustic environment in a factory, it is desirable to have information on several features of the noise sources:

Position of sources (possibly several in one machine)
Position of personnel relative to sources
Sound-pressure level at operators' positions of each machine
Sound-pressure level of each machine or process.

The definition of the sound-pressure level or power level emitted by a machine may not be entirely straightforward. Few machines in the average factory emit steady noise. Many machines emit noise only intermittently or impulsively. Many machines are not inherently noisy. They only become noisy as a result of interaction between the machine, its tooling and the raw material or product. A good example of this is the punch press. Most of these will run quietly and sweetly without tooling. The press manufacturers do not make tooling and for the standard types of press may not have exact details of how each one is to be used. Hence it is not much use going to a press manufacturer and asking for detailed information on press noise when in use. Metal-cutting machines also pose similar difficulties. The noise emitted by a lathe is quite different when running off-load than when on-load, when an appreciable amount of noise may be emitted by the object being turned. An even more extreme case is that of hand-held tools, particularly those with a reciprocating action. Most of the noise emitted by them when in use is secondary radiation from the structure or object which is being operated on.

The noise of non steady sources of noise has to be described by means of the equivalent continuous sound level (L_{eq}) over a typical operation or operating cycle. The L_{eq}s from a number of machines can be added together as if they were steady noise levels.

3. LAYOUT

Machine and process layout may often be outside the scope of a building designer, but there are many ways in which he may interact with the factory management and the machine layout designers. Factory layout can have a major effect on the noise exposure of employees. There is little excuse for excluding noise considerations from the normal design parameters; it should be considered along with process design, for instance, for factories of the 1980s.

In many industrial buildings the noise exposure of many employees is higher than it need be, as a result of the unfortunate juxtaposition of noisy machines and quiet manual tasks. Two examples of this are:

1. Lines of high-speed presses disgorging finished components to packers positioned one or two metres from the press tools. The people do nothing more noisy than packing, say, metal foil containers into cardboard boxes, yet work in an L_{eq} of 95 dB(A) plus. If consideration to the noise problem is given at the right time, a screen wall could easily be erected between the presses and the operators, with noise-attenuating delivery chutes passing through the wall. There should be little difficulty in producing a packers' working environment of 80 dB(A) or less by this method.
2. Commonly encountered is the intermixing of light assembly tasks with noisy operations such as presswork.

There is an understandable desire to keep layout as flexible as possible and install a wide variety of functions in one large open space. There is obvious justification for this in a car-assembly plant, for instance, but in other industries rather more thought needs to be given to the concept than has hitherto been the case. To improve the productivity of the shipbuilding industry, a number of yards have decided to roof in very large spaces, big enough to contain several medium-sized ships.

This is logical for productivity but can increase the noise dose received by the shipyard workers, because a number of occupations previously carried out in outdoor conditions are now in a reverberant indoor space. The situation tends to become worse if all the operations leading up to the construction of a ship are enclosed. In more than one case, the work of a plate shop, forge, machine shop, inspection area and instrument shop have been brought together under one roof from many diverse locations. This is potentially disastrous from the noise point of view, because plate-shop operations and structural sub-assembly are quite incompatible with quiet activities such as instrument work.

Figure 11.1 Reduction of sound level with distance from noise source in room

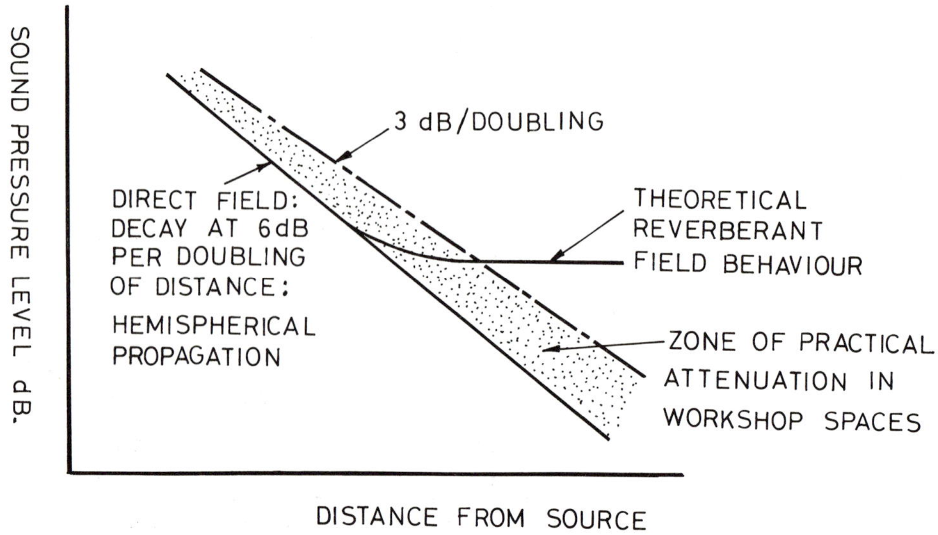

4. SOUND PROPAGATION

In other chapters the behaviour of sound in small and large rooms has been discussed. The classical acoustic model of a room often does not apply to real factory spaces, which may be too big, or too tall and narrow, or too long and flat (see Figure 11.1). As a result, it is often not possible to find zones in the direct field of a noise source, in which the noise decays by 6 dB per doubling of distance. It may also be impossible to distinguish a reverberant field in factory spaces. In very large factory buildings sound levels may decay very much as if the sources were in the open air, except close to the walls, where their reflections will affect the sound field. It is as well to bear in mind that the calculation procedures for sound in rooms usually assume flat walls and ceilings. We have come across instances of large industrial buildings with roofs of part-circular cross-section. These can cause appreciable increases of noise, due to focussing of the reflected sound.

When considering the noise dose likely to be received by employees working on the shop floor, it should be remembered that in general it consists of the noise related to their own occupation, which in most cases cannot be influenced by the building designer, plus the noise received from all the other activity around, which may be influenced by factory layout or building design.

So far we have only considered what may be called workshop noise, but noise in other parts of the factory must also be considered under the heading of layout. It should be obvious that it is unwise to site offices immediately adjacent to, or directly opening on to, noisy workshop areas. Less obvious to those unfamiliar with noise problems may be the need for great care in the siting of compressor rooms and HVAC plant-rooms in relation to sensitive spaces. There is the classic mistake of siting the air-handling plant on the roof, directly above the MD's pent-house office – an almost certain way of lowering the esteem of the architect in the eyes of the MD! For noise and vibration reasons the place for a plant-room is in the basement, where compressors and fans can be installed on good solid concrete bases. Even that is no guarantee of freedom from trouble. We know of one case where a brand-new expensively equipped factory medical centre was built above the main plant-room (see Figure 11.2). That may not have been too serious if two major mistakes had not been made. The first was that all the pipework was connected to the underside of the medical-centre floor slab, so if you wanted to know which compressors were turned on at any time you could always visit the medical centre rather than bother to go down two flights of stairs to check. The second mistake was that the plant-room had large double doors to allow access for machinery but very little natural lighting or ventilation. As a result, the doors were often left open. That, also, would not have mattered if there had not been a high brick wall opposite, from which all the plant-room noise was reflected back into the medical-centre windows.

Figure 11.2 Paths for transmission of noise from plant to occupied spaces

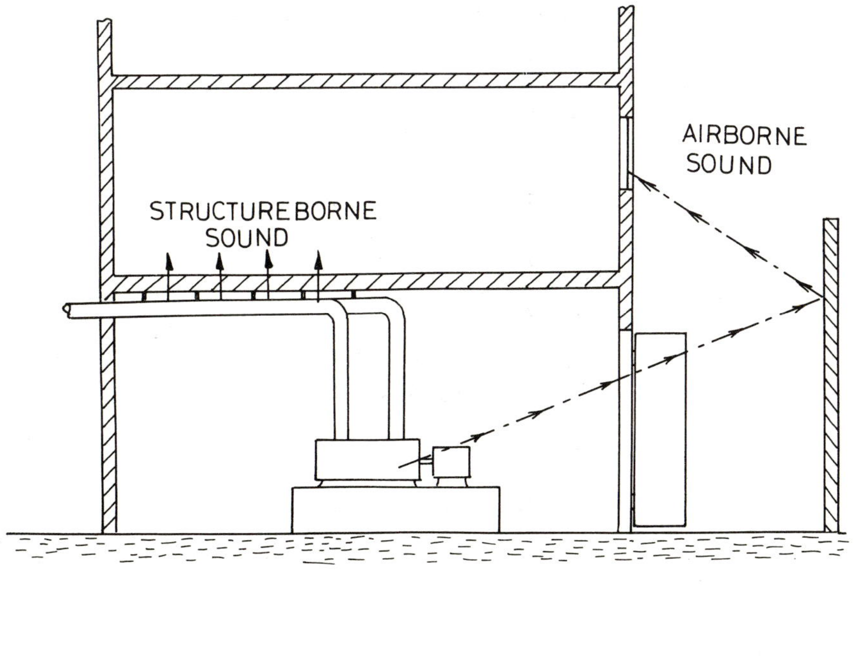

Figure 11.3 Types of silencer: (a) Absorption, (b) Reactive

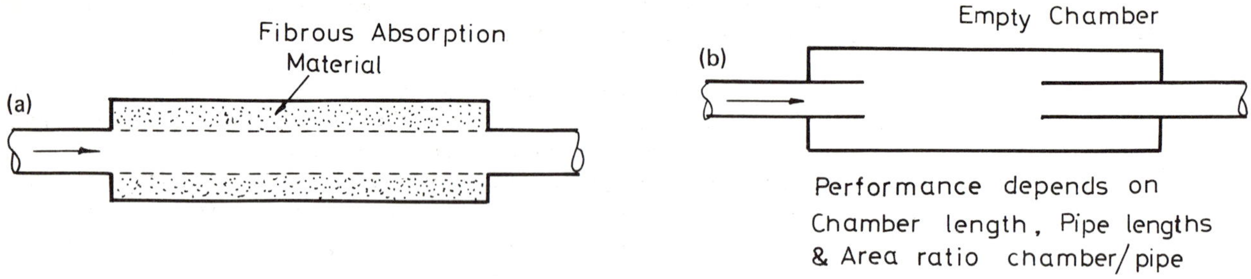

Another example of what some might call bad luck and others incompetence is the case of the drawing office above the compressor room of a large factory. The factory compressed-air supply was produced by several large low-speed reciprocating compressors drawing air from inside the compressor room. The compressors had been fitted with normal inlet silencers, which reduced the dB(A) level in the room but were quite incapable of dealing with the main pumping frequency of about 20 Hz. As a result, the whole drawing office reverberated at 20 Hz. The noise, being so low in frequency, could be felt as much as heard and caused severe distress to the draughtsmen. It also produced visible shaking of various fixtures and fittings. The installation of large reactive silencers (see Figure 11.3) able to deal with the 20 Hz excitation would have produced a solution to the problem, but in this case there had been such a degree of dissatisfaction among the DO staff that the management decided that relocation of the compressors and replacement with higher-speed machines was justified, because they had been considering it for several other reasons, too.

This problem highlights one of the dangers of using the conventional definitions of noise in cases of low-frequency noise in buildings. It is quite common for dB(A), NR or NC levels to fall within accepted criteria for office noise, for instance, but for the noise to be quite unacceptable to the occupants because the spectrum shape is severely unbalanced at the lower end. More detailed analysis of the problem is then required. It is easy for these low-frequency excitations to tune into transverse or longitudinal modes of rooms and cause increased discomfort for the occupants. For the room resonance to be excited, the length, width or height of the room needs to be close to a multiple of half a wavelength of the exciting sound.

5. NOISE CONTROL OPTIONS AVAILABLE TO THE BUILDING DESIGNER

Apart from the location of sources, the building designer has two major noise-control options open to him:

1. Sound absorption
2. Barriers and enclosures.

5.1 Absorption

One normally thinks of absorption as being provided by acoustic ceilings and other intentionally installed absorption materials, but it is as well to remember that almost anything in a workshop adds absorption. Machines, work in progress, stores, people, etc, can all have a significant effect, particularly in a building which otherwise has very little absorption in it. The Machine Tool Industry Research Association has evaluated the 'natural' absorption effects in a number of workshops, using reverberation time measurements [2]. They found that the average absorption coefficient (based on dB(A) measurements) of a number of workshops was about 0.2. It may be safer to use a rather lower value of 0.1 to 0.15 for design purposes, because the MTIRA tests used pistol-shot excitation, which has a somewhat different frequency spectrum from most machines.

Those unskilled in noise control often regard the addition of absorption as the universal cure for all factory noise problems. It is not. Wrongly used, it is an expensive white elephant.

An important rule to remember is that the addition of absorption in a space does not affect the direct sound received by a listener; it can only affect the reflected sound. Thus, the addition of absorption will only reduce noise significantly at an operative's position if the reflected sound is within about 6 dB of the direct sound. If the reflected sound level is more than 10 dB below the direct sound, the addition of absorption will certainly not reduce total sound by a measurable amount.

Using the conventional room acoustics equations, the effect of additional absorption on the noise level in the reverberant field can be estimated, but we have already seen that true reverberant fields do not often exist in real life, so what do we do then? Friberg [3] developed a method of assessing the acoustic properties of large industrial rooms or workshops based on results from a large number of practical measurements. His method is generalised and simple to apply but seems to give good results. It enables the effects of increasing the level of absorption in a workshop to be assessed.

In industrial buildings, absorption is usually installed at ceiling level. It is most likely to be successful in buildings which are long, or wide, and low; in which there are large numbers of relatively low-power noise sources; or in which there is a large distance between source and receiver. It is less likely to be successful in very large and very high rooms, where reflected sound is, anyway, of low level.

Ceiling absorption may be in the form of a surface covering such as ceiling tiles, or hanging slabs or shapes of absorbent known as functional absorbers. The latter are often simpler to install by hanging on tensioned wires, and can be fitted without interference with lighting, ventilation or fire-extinguishing fittings. Per unit area of absorption, functional absorbers are usually highly effective. They can achieve an apparent absorption coefficient of well over unity.

Surface coverings of absorption materials can potentially be made to serve the dual purpose of acoustic absorption and thermal insulation, which can cushion the shock of the rather high cost of materials and installation.

Absorptive materials can cause some hazards in the industrial environment. Some materials may not be adequately fireproof, may allow fibres to be shed, or may allow fluids or particles to be absorbed. All these factors must be considered in the choice of absorption material. The ability to clean the material may also be necessary. In recent years an increasing number of surface-coated materials have become available. These overcome many of the disadvantages of fibrous and porous materials but at the same time allow most of the acoustic performance of the materials to be maintained.

The characteristics of all absorption materials are frequency-dependent and selection of materials may have to be based on a careful consideration of the spectrum of the noise in the enclosed space. The thickness of the absorption material or the airspace behind it usually has a profound effect on the frequency characteristics. Thick materials and deep airspaces give improved low-frequency absorption.

When very low-frequency absorption characteristics are necessary, for instance in power station turbine halls which are affected by high noise levels at 100 Hz, special-purpose absorbers are needed. These usually consist of tuned cavities of one type or another.

5.2 Barriers and enclosures

Design techniques for enclosures are well known. As far as the building designer is concerned, enclosures will usually take the form of separate rooms.

Design techniques for indoor barriers are less well known. A useful and simple-to-apply technique has been proposed by Aishton *et al.* [4]. This is based on a modification of the classical room-acoustics approach and thus suffers from the disadvantages of that approach, but nevertheless gives a useful guide. In very large indoor spaces where sound dies away at approaching 6 dB per doubling of distance from a source outdoor barrier calculation techniques can be used as described by Maekawa [5].

Important points to remember about barriers are that out of doors they are most effective when placed close to a source or close to a receiver, but indoors they are effective only when close to the source. They are useless in the reverberant field, because in that region the direct sound is of insignificant level.

6. NOISE EMISSION FROM FACTORY PREMISES

It is rarely admissible to consider only the noise inside factory premises. The noise from the premises is liable to escape outside to the neighbours, and activities inside the premises may require outside activities which cause noise in the neighbourhood.

Figure 11.4 Requirements for successful control of noise emanating from a building

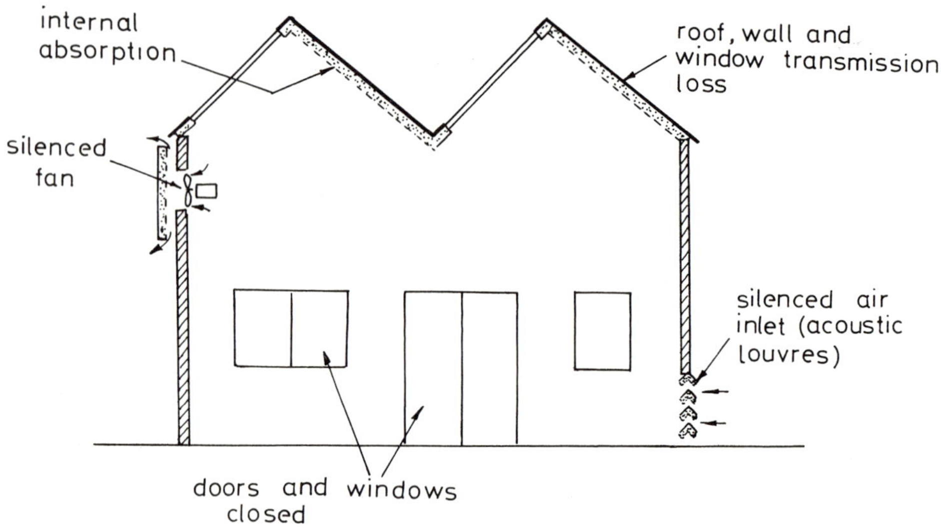

To minimise noise escaping outside, the following should be borne in mind (see Figure 11.4):

1. Adequate wall and roof transmission loss is required.
2. Internal absorption is beneficial if it reduces the sound-pressure level inside the walls or roof.
3. Ventilation openings are likely to be required. Unless dealt with by adequate silencing, they will let out noise. Ventilation fans will also make noise outside and must be selected or silenced to be adequately quiet.
4. Open doors let out much noise. The building, as a total system, must be designed in such a way that doors do not need to be left open.
5. Care should be taken about outside activities. Problems often occur because of:
   (a) Traffic movements. (Fork-lift trucks rattle and have noisy engines, and are often left idling for long periods. This is also true of lorries, which may also make noise loading or discharging their cargo.)
   (b) Scrap disposal. (This is often a noisy process, requiring care in siting and the erection of adequate screening, if necessary.)
   (c) Plant-rooms and compressor rooms. (Particular care is needed with apparently passive objects such as air receivers, which can be very efficient noise radiators; also, with auxiliary generator set exhausts, and compressors used for temporary works. External steam piping can be noisy and package boilers are notorious for generating annoying low-frequency noise.)

REFERENCES

1. *Code of Practice for reducing the exposure of employed persons to noise*, HMSO, London (1972).
2. *Effect of the test environment on the measured noise from machine tools*, Machine Tool Industry Research Association, Report No. 75.
3. Friberg, R. 'Noise reduction in industry halls obtained by acoustical treatment of ceilings and walls', *Noise Control and Vibration Reduction*, (March 1975).
4. Aishton, T.H., Moran, M.J. and Faulker, L.L. 'Sound pressure level prediction in large rooms containing barriers', *Applied Acoustics*, **11**, (1) (January 1978).
5. Maekawa, Z. 'Environmental sound propagation', Invited paper, 8th ICA, London (1974).

12 Design of recent auditoria

Michael Barron

SYNOPSIS

The principal subjective and objective characteristics which contribute to good acoustics are discussed in this chapter. The implications for concert-hall shape are drawn and the characteristics of recent British auditoria are compared with these criteria. The requirements for speech and, in particular, thrust-type stage theatres are also discussed.

1. INTRODUCTION

The form of auditorium which has been most intensely studied is the concert hall. As the concert hall also constitutes the most complex situation, it is convenient to discuss this first.

2. REVERBERATION TIME

The reverberation time remains, justifiably, the primary consideration for the design of an auditorium. The optimum value is considered to be a function of programme, recommended values ranging from 1.8–2.2 sec for symphony music to 0.8–1.1 sec for speech and drama.

The reverberation time is generally independent of position in the hall and is thus a characteristic of the hall as a whole, which makes it a convenient acoustic label. In conditions of adequate diffusion, the reverberation time of auditoria can be predicted with reasonable accuracy.

The reverberation time (RT) is a measure of the decay rate of the later sound. Though the subjective significance of this later sound has been seriously questioned, in conditions of adequate diffusion the RT determines several quantities which are now considered more important.

3. IMPORTANT CHARACTERISTICS OF THE EARLY SOUND

The fact that listening conditions vary within an auditorium and that halls with the same reverberation time generally 'sound different' implies that other factors are also relevant. It has now been established that the concert-hall experience is multi-dimensional and that no ideal concert hall can exist since the various desirable qualities are somewhat contradictory. The greatest concert halls are characterised by particular qualities, such as the reputed clarity in the Royal Festival Hall, London, and the warmth and reverberance of the Musikvereinsaal, Vienna. It is useful to consider the passage of a sound wave as it radiates from a point on the stage. The listener first receives the direct sound from the source, followed by a series of reflections off the walls and ceiling. In later time periods reflections arrive with diminished amplitude, which have been reflected by a series of surfaces, though the density of these reflections increases with time, such that after about 100 msec after the direct sound the situation is one of a reverberant field. Various characteristics of the early sound are now considered relevant.

Principally, three characteristics appear important for good acoustics: the correct balance between clarity and reverberance, the spatial impression associated with early lateral reflections and the loudness at the listening position. The perceived degree of clarity appears to be related to the ratio of early to late energy arriving at the listener. The early energy includes the direct sound as well as early reflection, and the limiting time period is now normally taken as 80 msec. This ratio will always be high for seat positions close to the stage, where one hears with great clarity. At more remote seats the value is often close to that associated with a linear decay at a rate determined by the reverberation time. However, this is dependent on good diffusion and uniform conditions, which become significantly more difficult to achieve in larger auditoria.

Regarding the second characteristic, there is now considerable evidence that strong early lateral reflections produce a highly desirable spatial effect, giving a sensation of envelopment by the sound, particularly in the *forte* passages. The effect is produced by prominent reflections from the side walls, which are clearly more significant in narrow halls. As an objective measure, the early lateral energy fraction is relevant.

Finally, it appears that the absolute loudness is important: at least in large halls, louder listening conditions appear consistently to be preferred. Since loudness is generally less in larger halls, this is one of several aspects which suffers in the largest concert halls. Seat positions remote under balconies are also likely to suffer from inadequate loudness.

4. DIFFUSION

The need for adequate diffusion has been a preoccupation of acousticians for several generations. The problem is, basically, that the vast majority of the absorption, in the form of the audience, is concentrated on only one of the six surfaces (of a rectangular room). This is not a condition in which one expects a uniform distribution of reverberant sound energy arriving at the listener uniformly from all directions. In the past the provision of diffusing elements on relevant surfaces, particularly the ceiling, has been considered the solution to this problem. It is possible that, on occasions, a more radical if less apparent solution to the problem is required, and that, for adequate diffusion, reflective surfaces, such as balcony fronts, are required distributed among the (absorbing) audience area, as well as a highly structured ceiling 'opposite the audience area'. These reflective surfaces are likely to be much more efficient as diffusers than the surface modelling up to a depth of 0.6 m normally prescribed.

A condition of adequate diffusion is a factor particularly difficult to predict but one which acoustic models are well suited to determine.

5. THE CURRENT APPROACH TO CONCERT HALL DESIGN

The approach to concert hall design found in several books of about twenty years ago can be roughly summarised as follows: the volume of the hall should be selected for a suitable reverberation time; adequate diffusion should be included on the wall and ceiling surfaces; balconies should be carefully designed to avoid excessive screening; surfaces liable to produce echoes are to be treated, etc. The importance of room shape was appreciated but apart from comments relating to rectangular halls there was little coherent advice.

Research over the last twenty years has concentrated on the subjective characteristics for good acoustics. Although the results of these studies have varied and been sometimes vague, a consensus view can be drawn in the form of objective measures which are considered important. Three of these measures have been mentioned above. However, whereas reverberation time is more or less constant throughout a space and, with diffuse conditions, can be calculated, these other measures vary over the seating area and cannot be predicted from drawings.

This complication can be overcome by making measurements in acoustic models. Traditionally, model scales of 1:8 or 1:10 have been used. An acoustic model has the property of reproducing in miniature the acoustic sound field: the time scale is reduced and the frequency increased by the scale factor. Air absorbs high frequency sound, so that for optimum modelling tests are conducted in dehumidified atmospheres. Subjective testing by playing speech and music through models of 1:8 or 1:10 can also be made. A recent development is the testing of small models, with scale factors such as 1:50. This has advantages of economy and flexibility with the model at the expense of a slight loss of precision.

Models allow the significance of hall shape and the effectiveness of surface treatment to be measured before there is full commitment to the design. If faults are found, changes can be made and tested at little expense. Acoustic models considerably increase the knowledge of behaviour of a space relative to what can be gained from drawings alone. Although objective measures other than reverberation time and the use of acoustic models are not at present universally accepted, this is likely to take place over the next decade. The significance of different hall shapes remains an area capable of more study, yet some general remarks can be made.

6. PLAN SHAPES IN CONCERT HALLS

The world's concert halls with the best acoustic reputation are rectangular in plan and roughly square in section. They also contain highly decorated wall surfaces, and balconies along the side walls. The two best-known examples of such so-called classical halls are the Musikvereinsaal already mentioned and the Boston Symphony Hall, USA. The St Andrews Hall, Glasgow, was also similar in design and highly reputed. The supposed reasons for their success are their satisfactory RT, their high degree of diffusion and the presence of surfaces capable of providing strong lateral reflections.

Unfortunately, the rectangular plan form cannot be simply scaled up for larger audiences. Principally, this is due to sight-line problems and the need to restrict the width of the stage. An interesting solution to the stage problem is to be found in the Royal Festival Hall, in which at the lower level the width tapers somewhat towards the stage end. A more logical solution to this problem is the use of the fan-shape in plan for the full height. This shape, however, has a poor reputation, probably due to poor diffusion and poor early lateral reflection conditions. A compromise solution which has been used successfully in recent years is the elongated hexagon (e.g., De Doelan, Rotterdam, and the Sydney Opera House Concert Hall).

7. CONCERT HALL SHAPES IN LONG SECTION

Two comments deserve mention with regard to the long section. Firstly, low ceilings over (remote) sections of the audience appear undesirable. Secondly, the trend pursued in the 1950s of ceiling and orchestral reflector design to project sound to the rear of the audience is now out of vogue. It appears that satisfactory listening conditions occur without this projection of early sound, even in remote seats, and such methods can introduce undesirable subjective effects.

8. RECENT CONCERT HALLS

8.1 Royal Festival Hall, London

A convenient starting point for the discussion of concert halls in Britain is the Royal Festival Hall, completed in time for the Festival of Britain in 1951. It was the first full-size concert hall to be built in Europe for a long while and represented the embodiment of acoustic thinking at the time. It seats an audience of 3000; to accommodate this number a plan width of 32 m was necessary, wider than in classical halls, compared with widths of 23 m for the Boston Symphony Hall and 18 m for the Musikvereinsaal. However, the plan shape is basically rectangular. The width at stage level tapers towards the stage, as already mentioned, with seating at a higher level running round the stage to join up with choir seating behind the stage. Boxes occupy a substantial proportion of the parallel side walls.

While the plan shape is a development of the classical rectangular plan, the long section is not inspired by the classical designs but bears some features of the 'directed sound'-type halls mentioned by Brian Day in Chapter 10. The concern that listening conditions might be poor at seats in the rear of the hall resulted in a reflector being placed above the stage and the ceiling being somewhat profiled to direct reflections to these distant seats. The combination of these elements certainly maintains satisfactory listening conditions at the back of the Grand Tier, but it now seems probable that this would have been achieved without these specific measures.

The Festival Hall has since its opening established a good reputation. The quality of these good acoustics does not, however, correspond with that attributed to the world's greatest halls. The Royal Festival Hall is noted for its high degree of clarity, described by some as clinical but liked by many musicians who cherish the ability to listen to indivi-

Figure 12.1 Royal Festival Hall, London (a) plan and longitudinal section, (b) photograph (copyright M. Barron)

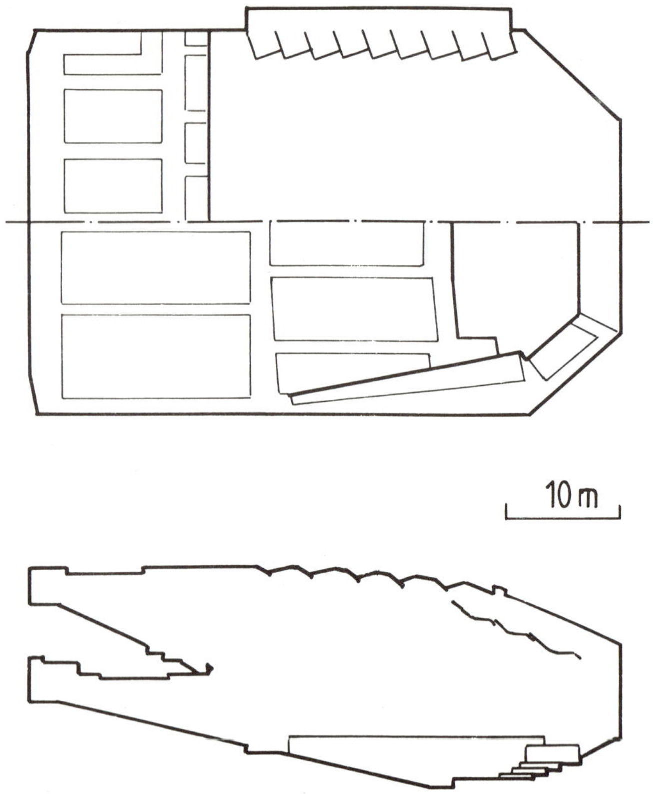

(a)

(b)

dual groups of instruments if not also individual performers. There are several reasons for this clarity: principally, the ceiling and reflector design already mentioned, and the relatively short reverberation time of 1.7 sec. This reverberation time is, in fact, shorter than planned, apparently due to the audience absorbing more than figures available at the time would have suggested. The suspended ceiling was, also, not constructed as planned, which caused considerably more absorption than one would have liked, particularly at low frequencies.

To remedy this short low-frequency reverberation time, an electronic system capable of artificially extending the reverberation time was developed, known as 'assisted resonance'. In the Royal Festival Hall, this has been widely considered as providing a worthwhile improvement. The 'assisted resonance' system has been subsequently incorporated into the design of halls to provide a variable acoustics facility, an example of which is discussed below.

8.2 Queen Elizabeth Hall, London

The Queen Elizabeth Hall, also on London's South Bank, could hardly present a more obvious contrast in architectural style to its neighbour. It is a hall designed for chamber music or solo performance, and seats 1100 people. It has a reverberation time which is, in fact, longer than that of the Royal Festival Hall at 2.0 sec. As it is constructed as a solid concrete shell, there is a need for low-frequency absorption to avoid excessive low-frequency reverberation times. This has been achieved by Helmholtz resonators on the side walls. To cover a range of frequencies, there are openings of various sizes for individual resonators; a piece of porous material placed in the neck provides the necessary damping.

Figure 12.2 Queen Elizabeth Hall, London – plan and longitudinal section

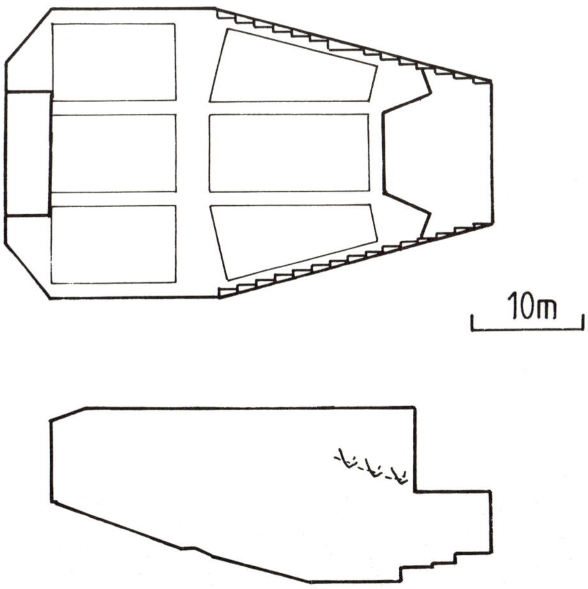

The overall acoustic reputation of the Queen Elizabeth Hall is very good. An examination of the hall plan indicates how, in this case, the problem of accommodating a narrow stage has been tackled. Instead of using a simple fan shape, the side walls, where they contain the resonators, are stepped. Another interesting feature of this hall is the reflector above the stage. This has been designed to be movable so that it can be placed either to provide a reflection onto the audience or not. At the time of the opening of the hall several subjective tests were conducted in the hall to determine whether or not such a ceiling reflection was desirable. Following the results of these tests, it is only for solo piano performances that the reflector is now used. This provides evidence in support of the current trend in concert-hall design to use reflectors above the stage only to reflect sound back down to the musicians.

8.3 The Maltings, Snape, Suffolk

The Maltings Concert Hall at Snape, near Aldeburgh, was completed in the same year as the Queen Elizabeth Hall, 1967. Its acoustics have been widely acclaimed; clearly, the reasons for this merit investigation.

Figure 12.3 The Maltings, Snape, Suffolk (a) cross-section, (b) Photograph (copyright The Architects' Journal, 13 September 1967)

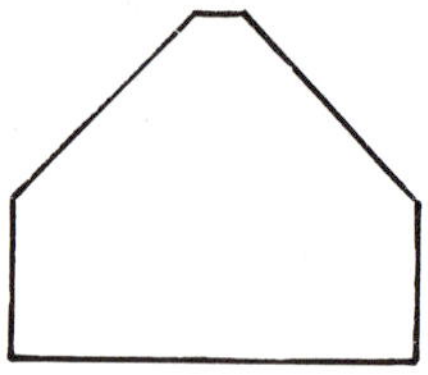

(a)

(b)

The first aspect to note is that the seating capacity of only 824 is relatively small by concert hall standards. As is widely appreciated, the problems of acoustic interiors increase dramatically with increasing size. The plan shape corresponds very closely to that of the classical halls, with a width of only 18 m. This provides strong early lateral reflections, which are considered so desirable for the best acoustics.

The unusual feature of this hall is the roof shape, which is gabled rather than flat. It is, however, difficult to be specific about the virtues or disadvantages of a gable-type roof. It has been suggested that such a roof will provide additional desirable early lateral reflections but a study assuming geometrical reflection in this particular cross-section showed that a flat-ceilinged hall of equivalent volume behaves equally well in this respect. Nevertheless, it is possible that a gabled roof may help create a high degree of diffusion. From a simple-minded point of view, sound cannot travel round as in a rectangular space confined in a horizontal plane, without hitting any of the audience. Gabled roofs deserve further study.

The Maltings is also used for opera, using a pit in front of the stage. It works well for this function also, though again the modest size probably helps considerably.

8.4 The Hexagon, Reading

The Hexagon seats 1500 people and was opened in the autumn of 1977. It was designed as a multi-purpose space, with the 'assisted resonance' system to prolong the reverberation time for music use. 'Assisted resonance' has already been mentioned above in connection with the Royal Festival Hall, where it was used to correct a design deficiency. In the Hexagon it is employed to provide variable acoustics. By extending the reverberation time, the hall could be built with the smaller volume suitable for speech. In principle, one selects the hall volume to provide a reverberation time of around 1 sec, suitable for speech, and the assisted resonance system roughly doubles the reverberation time so that it is suitable for music.

In design, the hall is basically on two levels: the stalls floor is nominally flat, containing loose seating and at the rear bleacher seating; the balcony areas are subdivided into seating blocks but are otherwise continuous, choir seating continuing at the same level round the back of the stage.

The stage is used for an amazing range of events, which include boxing, theatre (with proscenium), ballet, concerts, with all-night jazz sessions and beer festivals as more extreme uses. As a hall for a large conurbation it has proved very successful. Acoustically it appears successful too. It works well for theatre, and well for small musical groups. The prominent ceiling panels clearly assist intelligibility of speech. The success for orchestral music is obviously dominated by the success of the assisted resonance system.

The natural mid-frequency reverberation time is about 1.25 sec; this can be raised to 1.7–1.8 sec with the assisted resonance system. With orchestral music, the change is apparently valuable. If it does have shortcomings, the sound energy from the system is a little uneven, being excessively evident in the upper balcony regions and probably too quiet in the stalls. This is a consequence of the microphone and loudspeaker placement in the auditorium.

The most severe criticism of the Hexagon has been of the unsuitability of the stage facilities for theatre. The Hexagon has no fly tower and scene changing mid-performance is almost impossible.

8.5 Eden Court Theatre, Inverness

This recent hall has a different solution to the problem of multi-purpose use and makes interesting comparison. Eden Court has a fly tower and for orchestral performances a stage enclosure is used. The design is similar to that of traditional opera houses, with seats arranged in tiers following a horseshoe shape in plan. Due to its geographical position this hall is not familiar to many but the reverberation time does deserve mention. At Eden Court the reverberation time is not variable and has been chosen to be intermediate between requirements for speech and music at 1.45 sec. This appears to cause some minor problems for speech but only at a few seats.

Figure 12.4 The Hexagon, Reading (a) plan and longitudinal section, (b) photograph (copyright M. Barron)

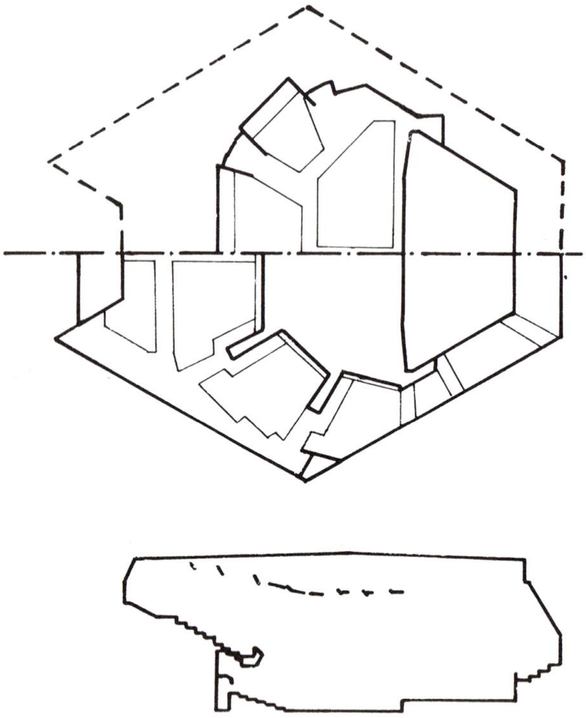

(a)

(b)

Figure 12.5 The Crucible, Sheffield (a) plan and cross-section, (b) photograph (copyright The Architectural Review, February 1972)

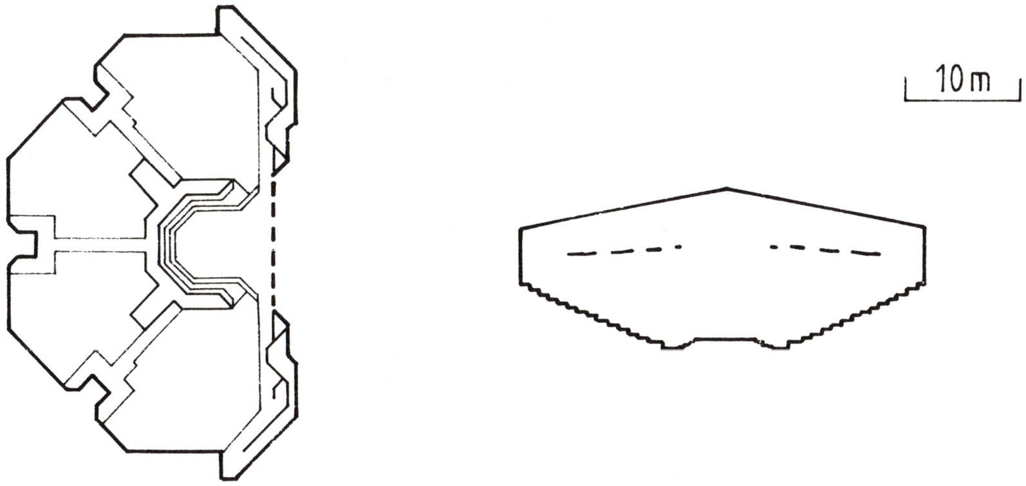

(a)

(b)

9. THEATRE DESIGN

The acoustic requirements for theatres are more easily definable. The aim in room acoustics is generally two-fold: firstly, to transmit information and, secondly, to provide a suitable environment in which this information is heard. For music, it is the environment which probably receives most attention. For speech, it is the information which is important and the ease with which it can be heard. The acoustic requirements for intelligible speech are also more easily quantified. In practice, in the conventional proscenium-type theatre holding about 1000 people the acoustics generally look after themselves if the distance to the farthest seat is about 20 m.

A different theatre form has, however, emerged during the past twenty years, generally known as the thrust-type stage theatre, pioneered by Tyrone Guthrie. This theatre form presents a considerable challenge acoustically, for the basic reason that actors are frequently facing away from members of the audience. The problem arises because the human speaker is directional, he projects more sound energy forwards, particularly at high frequencies which are critical for intelligibility. With a thrust-type stage theatre there are frequently people directly behind an actor, and in full theatre-in-the-round there will always be some of the audience behind the speaker.

The acoustic implications of this are that:

(a) The reverberation time must be kept short.

(b) The direct sound must be supplemented by strong early reflections at the more remote seats.

(c) The maximum distance for seating from the stage will be smaller than in proscenium-type theatres.

10. RECENT THEATRES

10.1 The Crucible Theatre, Sheffield

This theatre, completed in 1971, is the most recent of a series with thrust-type stage. It holds an audience of 1000. The design, in plan, is a major segment of an octagon, and the seating is on a single level, with a substantial rake. The result is not unlike one end of a Roman arena.

In a theatre of this shape, the only really valuable surface available for providing reflections is the suspended ceiling and with this there is likely to be a conflict of interest between the lighting and acoustics. In the Crucible a series of panels covering most of the seating area has been carefully designed to follow the line of the stage front. The panels provide good acoustic reflections and maintain good intelligibility at the furthest seats.

10.2 Royal Exchange Theatre, Manchester

Completed in 1976, this theatre seats 700 people in true theatre-in-the-round. It is housed in the middle of the now disused Royal Exchange building. For structural reasons, lightweight materials had to be used and the structure is partly supported from steel trusses built into four central stone columns.

The principal acoustic features of the design are that the plan is seven-sided to avoid parallel surfaces which might produce colourations or flutter echoes. The audience is seated on three levels and no-one is, in fact, more than thirteen metres from the edge of the acting area. This, together with the short reverberation time of about 0.8 sec, provides good listening conditions even when actors are facing away.

10.3 Olivier Theatre, National Theatre, London

Also opened in 1976, the Olivier Theatre is the most interesting and novel of the three auditoria in the National Theatre. The other two theatres in the complex, the Lyttelton and the Cottesloe, are smaller and conventional. The Olivier Theatre seats 1160, on two main raked tiers. In plan it consists of a circular stage centred close to the apex of a 90° fan. The furthest seat is about 22 m from the stage front, i.e., typical for proscenium theatres. There is, however, no proscenium as such, the splay side walls continuing some way into the fly tower and tilted panels above the stage front concealing lighting etc. The form of the stage allows considerable flexibility: for each production the stage set or black curtains define the acting area. The plan form is a novel variant of the thrust-type stage but it has full flying facilities without a proscenium.

Figure 12.6 Royal Exchange Theatre, Manchester (a) plan and section, (b) photograph (copyright The Architectural Review, December 1976)

(a)

(b)

One surprising characteristic of the theatre is its substantial volume. For the auditorium itself, without the fly tower, the volume is 13 500 m^3, which is, in fact, larger than the volume of the Queen Elizabeth Hall. Both halls seat the same size of audience and in general one expects larger volumes for music than for speech. The reason that the reverberation time in the Olivier Theatre is not of the order of 2 sec is that the opening into the fly tower is large and this opening acts as a highly absorbent surface. The initial reverberation time in the theatre was, however, considered too long and has subsequently been reduced to 1.15 sec.

Early acoustic problems were associated, however, not so much with the reverberation time but with the early reflection situation. A series of panels at the rear of the balcony provides valuable strong reflections for the rearmost seats. At the front of the balcony, however, these reflectors no longer help and very little early reflected sound reaches these seats, resulting in poor intelligibility, particularly when actors face away (a common occurrence in this theatre). During 1977 this problem, together with some minor echo problems, was investigated by Sound Research Laboratories and in an acoustic model at Cambridge. A series of remedial measures was undertaken, which appears to have virtually solved the initial problems.

Figure 12.7 Olivier Theatre, National Theatre, London (copyright M. Barron)

11. THE FUTURE

It is hoped that this brief survey has given at least some pointers to what is now considered important for good acoustics. The research progress of recent years is gradually broadening the scope of factors considered important to internal acoustics. Sole dependence on the reverberation time criterion implied a specified volume but otherwise complete freedom with the shape. For music, the significance of room shape is gradually becoming apparent, with, for instance, an appreciation of the need for surfaces capable of providing early lateral reflections. Acoustic models are making valuable contributions, allowing much greater confidence to be gained in the design before construction. Models have helped with the design of two concert halls due for completion within the next three years, the Barbican Concert Hall, London, and the National Concert Hall of Wales, Cardiff.

BIBLIOGRAPHY

1. Parkin, P.H., Humphreys, H.R. and Cowell, J.R. *Acoustics, noise and buildings* (4th ed), Faber and Faber, London (1979).
2. Barron, M. 'Concert halls, opera houses and auditoria'. Ch. 2 in *Planning: buildings for administration, entertainment and recreation* (ed. Mills, E.D.), Newnes–Butterworth, London (1976).
3. Mackenzie, R. (Ed.). *Auditorium acoustics*, Applied Science Publishers, London (1975).
4. Kuttruff, H. *Room acoustics* (2nd ed), Applied Science Publishers, London (1979).
5. Reichardt, W. *Gute Akustik – aber wie?*, VEB Verlag Technik, Berlin (1979).
6. Jordan, V.L. *Acoustical design of concert halls and theatres*, Applied Science Publishers, London (1979).

Discussion

OPENING

Dr C.L.S. Gilford

Ladies and Gentlemen – I do feel honoured to have been asked by Dr Croome and the Construction Industry Conference Centre to open this conference on *Noise and the Design of Buildings and Services*. This is no ordinary technical meeting, not only on account of the high professional level of the contributors, but also because it is supported by four Institutions – the Royal Institute of British Architects, the Chartered Institution of Building Services, the Institute of Acoustics and the Institution of Public Health Engineers. This close interest by professional bodies reflects the widely interdisciplinary nature of the solutions to the problems of noise. These range from physiology to psychology and acoustics, mechanical and electrical engineering and even government administration and thermal design. The impact of noise on human society is made more intractable by several contributory factors: first, the extraordinary sensitivity and subtlety of the hearing mechanism and the minute power required to produce audible sound; second, the multiplicity of possible paths from the source to the hearer and the small losses along both air and solid paths, and the fact that the long wavelengths of sound make screening quite ineffective. Above all, the vast information-bearing potentialities of sound. Indeed, annoyance is mainly correlated with the hearer's feelings and attitudes towards the nature of the information in the sound and, except when comparing two closely related types of sound, hardly at all with the loudness or spectrum. The consequence of this is the plethora of different criteria of acceptability which have been devised to meet different classes of noise. Since individuals vary greatly in their attitude and sensitivity to noise, these criteria cannot, sometimes, be better than rules to prevent too large a proportion of the population from suffering distress.

This conference is divided into four natural sections. In the first we are dealing with human reactions to sound and with the question of criteria. The second concerns building services – there has been an extraordinary increase in the number and relative value of mechanical services over the last twenty years or so and indeed in some buildings, such as hospitals, they are now accounting for over half the total cost. As an example, ventilation or airconditioning equipment is now almost universal in commercial buildings. Noise in cities renders natural ventilation mainly impracticable and, in addition to that, open planning for offices and schools has produced a new application for airconditioning systems, that is, the direct provision of masking noise for distracting sounds which would otherwise be coming from other parts of the open space. On the whole, the industry has met the challenge with an increasing number of well understood components occupying very much less space than would have been possible a few years ago. Large ducts and roof-mounted fans have been replaced by small ones carrying air at high velocities to and from small air-handling units which are run at fast enough speeds to facilitate the damping of vibrations by simple mountings. Modern floating floors permit these units to be located in any spare space, even adjacent to sensitive areas. I am hoping that this session will tell of continuous innovation and refinement.

The third session will deal with design of buildings, and perhaps the major problem here is that of the design of economical sound insulation within buildings. Although the general characteristics of a particular construction are usually fairly well predictable, calculation of the transmission loss of even a simple cavity wall is seldom easy or accurate enough for routine use and the designer must have recourse to any available laboratory or field data. The number of possible variations in design is so great that the designer is often compelled to do a certain amount of rather doubtful estimation from data on very similar structures. There seems to be no central file of such information to which a designer can turn, in the end, for help, yet such data must surely exist and there is a great need for the pooling of results from case data by those who have to face these problems day by day. Further difficulties arise from the variability between nominally identical constructions. In this connection, there was an alarming revelation by the Building Research Establishment that a large number of dwellings constructed since the Building Regulations on sound insulation came into force failed to pass the tests.

Although a proportion of these failures could have been caused by design variations, not specifically anticipated in the Regulations, most were the consequence of sheer variability and the only cast-iron remedy is to overkill the design or by Draconian measures to enforce the Regulations by compulsory rebuilding of failures; the same applies to commercial buildings, of course. Many of the difficulties of prediction and maintenance of standards are most acute at low frequencies, where individual sensitivity may be most variable. The persistent complaints of low-frequency environmental noise which proves to be inaudible to anybody but the complainant point to the possibility that some of those concerned show anomolous behaviour to low-frequency low-level noise. I have met those cases not only with unexplained environmental noise, but also with specific identifiable low-frequency sounds at or below the average threshold of hearing. With the developing crises in fuel supply, the primary concern in building design is bound to be in inexpensive thermal design. In some respects, as with the use of cavities in walls, acoustic and thermal insulation requirements seem to be met by the same building measures. However, for thermal uses, very often the most economic type of building is of large blocks, and yet it is just in those types that sound transmission is a most obvious cause of distress.

The final session will be on room acoustics, chiefly in relation to large enclosures for work and entertainment. Perhaps the last chapter of all is a welcome reminder that, as I implied earlier, not all loud sounds are undesirable.

I am sure that we can look forward to an extremely interesting and stimulating conference.

SESSION 1

1.1 Dr H.C. Pinfold
University of Liverpool

Could I ask Dr Croome how arousal level is defined and how measured?

1.2 Dr D.J. Croome
University of Bath

The meaning of 'arousal level' is indicated by the attention continuum:

sleep	drowsiness	relaxation	attention	excitement	anxiety
LOW AROUSAL		MIDDLE AROUSAL			HIGH AROUSAL

One definition is that it is a measure of how awake the organism is and how ready it is to react; another that it indicates the inverse probability of the subject falling asleep. Arousal level is determined by the environment (i.e., changes, information rate, task being performed) and by internal human states (i.e., level of anxiety, drowsiness, alcohol, drugs). Research has obtained the intercorrelations shown below for the various factors of arousal.

Intercorrelations among the various factors of arousal-seeking tendency

Factors		*2*	*3*	*4*	*5*
1.	Arousal from: change	0.60	0.65	0.44	0.53
2.	unusual stimuli		0.53	0.57	0.45
3.	risk			0.41	0.40
4.	sensuality				0.27
5.	new environment				

Correlations in excess of 0.15 are significant at the 0.01 level.

It seems likely that arousal level increases when an increase in activation of the reticular system occurs.

Electroencephalographic recordings (EEG), variation in blood pressure and pupil dilation are examples of some of the physiological measures which are related to changes in arousal but more recent work reported concludes that this unitary description of arousal is too simple, and defines *electrocortical arousal* as that dependent on the reticular system, *behavioural arousal* as dependent on the hypothalamus and *autonomic arousal* as dependent on other parts of the nervous system. In general, all forms occur simultaneously, but are independent of one another.

The concept of arousal level does help to explain some of the personal differences in reaction to various environments. There is some evidence to suggest that females generally have a higher arousal level than males, and can hence better tolerate sensory deprivation. People who are extrovert, or those who tend to be neurotic and anxious, are often more arousal-seeking than introvert personalities. Introverts, however, are often very sensitive to variations in the arousal quality of environments.

When the arousal level is too low, the response of an organism to the environment is lethargic, insensitive and inert, but if too high, then the behaviour becomes tense and disorganised. Monotony, predictability and concordance tend to lower the arousal level, whereas sensory stimulation, novelty and conflict tend to raise it. For optimum arousal and best level of performance and least stress, moderate levels of these factors are required. Motivation can tip this fine balance towards achieving and maintaining optimum arousal, but factors such as lack of short- or long-term aims, an effort required beyond a person's capacity or willingness, can easily dim any motivation.

In physiological terms, arousal is thought to be concerned with the efficiency of information transfer between the cortex and the subcortical regions which house the arousal centres. Some work has attempted to relate the patterns of evoked cortical responses to arousal aspects of attention and there is much evidence supporting the view that the contingent negative variation (CNV) in these patterns is related to attention, and changes in amplitude and form of the CNV are closely related to moment-to-moment changes in conscious attention. The activation in the brain which accompanies arousal appears to be correlated with autonomic activity.

1.3 Dr H.C. Pinfold

It seems to me that in the chapter on human response to noise one thing that was missing was what people are accustomed to. To give an example of what I mean – we accept a party-wall reduction requirement of around 50 dB for privacy between dwellings. I would suspect that this arises from our culture having grown up in houses that have 9-in (225 mm) brick walls and that over a long period of time we have adjusted our verbal behaviour to that kind of protection. I think that if you went to Japan, where many people live with paper party walls, you would find that their party-wall requirement was quite different and, in fact, I was told by somebody who has made studies in Japan that the Japanese have an extraordinary way of not hearing things that they do not want to hear. I have seen this myself in Africa, where European whites customarily talk in a derogatory way about black people in the presence of their servants who happen to be hovering around the table serving dinner. I am told that in Japan it is quite customary to talk about people in their presence and for those people, unless being addressed, simply not to hear, or shall we say, simply not to listen.

1.4 Dr D.J. Croome

There have been some fascinating studies on attention switching, of which the 'cocktail party effect' is particularly well-known. There is no doubt that the information content of simultaneous conversations can switch the attention of one listener from one conversation to the other. Expectancy, too, is another example where the anticipated outcome of a situation has been programmed into the human system way back in our culture.

1.5 Prof J.B. Large
ISVR

What is being suggested here highlights the difference between laboratory data and what happens in real life. For example, I mentioned the difference in response between railway noise and aircraft noise and I am sure that part of this difference is a cultural one. We have lived with the railway for one hundred and fifty years or so, whereas aircraft noise is a new event and maybe we shall become socially habituated, but there is a big difference between the interpretation of laboratory data and the interpretation of the same phenomena in a real living situation, and I think one ought to distinguish between the two when thinking in terms of criteria, particularly design criteria.

1.6 Dr D.J. Croome

The degree of adaptation to environmental change is limited. Mentally, people in northern latitudes have never adapted completely to short hours of daylight, for example, although, physiologically speaking, they survive. We do not understand completely how noise affects the nervous system and an outer layer of social habituation may be no more than a survival mechanism, although the quality of our life as described by mood or nervous behaviour may have been permanently impaired.

1.7 P.R. Boyce
Electricity Council Research Centre

I wonder if the difference in the relationship between annoyance felt and equivalent continuous sound level for aircraft and railways could be related to the breakdown in reciprocity between noise level and duration occurring under different conditions for the two noise sources. L_{eq} is a dose-type measure and this approach may not hold equally well for both aircraft and railway noise. I would also be interested to know to what extent Professor Large considers an L_{eq}-type measure useful for defining a criterion to avoid sleep disturbance.

1.8 Prof J.B. Large

All these exposure indices, whether they are Noise and Number Index, equivalent continuous noise level (L_{eq}) or L_{10}, have the same components, which give a summation of individual events over a given time period and lead to a single definition. In the use of the equivalent continuous noise level there is some recognition of the duration of the individual events taken into account, but the business of a tail-off between duration and intensity is really, I think, not an important one, because the two are included independently of each other. Where you do have to be careful in interpreting the sort of data that I showed in my presentation is in the physical translation of noise and number index into L_{eq}, because, if you assume as many people do respond to just a constant value, you can end up with totally different results. One has really to break the noise down into its components again to convert it right back from the beginning arithmetically in terms of the individual noise characteristics.

As far as sleep is concerned, what I tried to explain is that it is a complex situation. Not only is it dependent upon the widely varying sleeping habits that we all have, but also the state in which we go to sleep is very important as to how our sleep is affected. For example, if you have a baby, you are half asleep, waiting to see what goes on in the next room, so that has to be taken into account, and therefore it is very difficult to define individual sleep characteristics with respect to noise intrusion. I feel that it is not satisfactory to use a simple single-index definition for noise and sleep disturbance. In other words, you are only going part of the way if you specify equivalent continuous noise level at

35, because in that environment one can have one or two very noisy events and still the L_{eq} will be perfectly acceptable. Therefore, my presumption is that one should not only specify the L_{eq}, because that gives you an indication of the overall noise climate, but also contain the individual noisy events, and here my assessment would be that one has to restrict them to a peak level of 50 dB and a peak rise in the background noise level of 10 dB, but I think you want all three of these in combination.

1.9 Dr D.J. Croome

Criticisms of the L_{10} criterion are, firstly, that in urban situations the L_{10} value can be very high due to traffic accelerating, decelerating and braking, whereas most data refer to freely moving traffic (e.g., I have measured peak noise levels of 92 dB(A) at 20 m from lorries braking at traffic lights); and secondly, people may have the quality of their sleep disturbed, but not necessarily be woken up, by night traffic. In Switzerland the emphasis for traffic noise criteria is placed on the desire to minimise sleep disturbance and the 1 per cent level L_1 has been advocated as a standard, which is even more difficult and expensive to measure than L_{10}, but it may be wiser to use such a standard.

1.10 R. Tuthill
DHSS

I would like to make some comments on what Professor Large was talking about and which follow on from the question of sleep; also, to request whether anyone has any valid information that may be of help to us. We had an interesting problem recently which we have not been able to resolve yet, and which is the converse situation, i.e., not to avoid waking people with noise, but how to ensure that people do wake up when a fire alarm goes off. In looking round, we have found very little data about this. What brought it to our attention was an incident one night at a training school of ours when someone, we do not know whether deliberately or not, set a fire alarm off about 1 am. It so happens that it had been installed recently and we measured sound levels of over 80 dB(A) in some of the bedrooms, but other rooms are in a very old, heavily constructed building with good acoustic insulation and therefore the sound levels are not quite as high in some bedrooms as seems necessary. We are worried about the low proportion of people that were actually woken up, and by analysing the response in relation to the known sound level in each room, it looks as though, even if you raise the noise level to something like 70 dB(A) in the bedrooms, there is still quite a large proportion of people who are not woken up at that level. Yet, to achieve levels beyond this, you have to ensure sound levels in the corridors so loud that they are liable to cause panic to some people.

1.11 Prof J.B. Large

This question of warning signals is a very important one and there is some research going on, not only in my department, but also in the United States. Very crudely, I think that, when referring to the population as a whole, one has to think in terms of a rise in the sound level above the background of about 30 dB. This means a peak sound level of at least 70–80 dB in the sleeping environment to make sure that about 95 per cent of the people wake up. Personal differences are very large. Sounds which startle are a very important aspect in the case of sleep, because people who are sensitive to noise can have a rather abrupt awakening, and perhaps the sleep warning signal should include a time effect so that there is a time delay before reaching the peak intensity. The whole matter of warning signals is important for fire, or hearing of police or ambulance sirens inside automobiles, and recently I was at a review of some American research at which they showed that you would only hear the police siren from a police car coming towards you at a distance of 4 metres. This is the distance at which you would start to hear the siren itself. Warning signals are a neglected area of acoustics and one where a great deal of research work needs to be done.

1.12 Dr D.J. Croome

Kryter, in his book *The effects of noise on man*, Academic Press (1970), suggests that the auditory threshold is gradually raised from about 30 dB in the lighter stages of sleep up to about 80 dB during deep sleep, hence making it increasingly difficult to wake people up by noise signals. Other research shows that older people are more likely to wake up than younger people. Frequency and change of frequency are important features of the warning signals – sound level is not the sole determinant.

1.13 B.A. Brown
Conder Group Services Ltd

I should like to ask Mr Taylor what problems he has had to overcome to ensure that doors do not reduce the sound properties of walls and what is the practical maximum reduction achieved for a single-leaf office door?

1.14 R. Taylor
Rupert Taylor & Partners

To answer the second part of the question first, it depends on what you consider as an office, and whether you permit special constructions to count as office doors. The most efficient single-leaf door that I have encountered was in a room constructed to house a typing pool, and achieved its effect by being artificially thick (not necessarily heavy) and fitted into a frame which contained a sound-absorbent lining. The frame formed a very, very thin sound attenuator, so no seal was necessary: the sound passing along the narrow gap between the door and the frame was attenuated by the lining of the frame. This seems to be a better approach than trying any sort of seal, because the only seal which is really effective is something like a phosphor-bronze strip but this makes it difficult to open the door when it is closed. I have not actually carried out measurements on this particular door but I would guess the average reduction over the 100–3200 Hz frequency range is about 38–40 dB.

1.15 P.H. Andrew
Kennedy & Donkin

I was interested to hear that when you wanted to increase the attenuation by the sound insulation of the ceiling tiles, you used steel plates to back them. Would not lead sheeting have been more effective under these circumstances and rather easier to fit?

1.16 R. Taylor

Yes, the choice of galvanised steel was for purely practical reasons and for cost reasons. The difference between lead and steel in that sort of situation where the sheet is well restrained and well damped by being applied to the back of an existing rigid structure would be very small and would depend on the relative masses of the two sheets. In many cases lead would be an easier thing to work with, but my solution was a question of economics and practicability.

1.17 J. Simpson
Department of Finance, Belfast

I have a question for Mr Taylor. I have been involved in shooting ranges and have been visiting them all over the country, and everyone seems to have different ideas for sound insulation. We shall be involved in a single shooting range and another project for a block of four shooting ranges. What would you advise for insulation? In the case of the four ranges, would you isolate each range from the other? Can you comment about annoyance to the public?

1.18 R. Taylor

There are three things to consider with a shooting range. The first is the hearing protection of the people firing the weapons and that will be unaffected by whatever you do to the range, because the ear is so close to the gun. They must wear proper hearing protection. The second is that the acoustics of the range will clearly affect the sound levels to which spectators nearby are exposed and, provided they are not within a metre or so of the person firing, then sound-absorbent treatment generally on all available surfaces will reduce the level. The third aspect is the case of noise escaping to other areas and causing annoyance, which is a classical sound-insulation exercise.

In the case of the four ranges, I do not see any need for isolating each range from the other. The only thing that really matters, from the internal point of view, is hearing damage. Unless there is a danger that the concentration of somebody taking aim will be upset by hearing shots from the next-door range, which I consider unlikely, I would not bother too much about cross-transmission between ranges. I would ensure that within each range the reverberant-sound level was as low as is reasonably possible, because it would be very difficult to ensure that everybody always wore hearing protection when just standing inside a range. One must therefore reduce sound levels for them to a safe level over their total period of exposure.

Annoyance to the public is a classic sound-insulation problem. There are no special problems other than those I have been talking about – air-tight construction, attenuated ventilation openings and ventilation ducts, and the application of classical engineering principles.

1.19 Prof J.B. Large

I was going to ask Mr Simpson a question, because I wanted to know what method of assessment you were using in order to come up with some judgement of what the sound levels from that range would be. [Mr Simpson replied that they had not used any method of assessment yet, as the shooting ranges had not been built.] The reason I am asking is that the Research Directorate of the EEC is now in the second year of a fairly extensive research programme looking at impulsive noise, which includes the noise from rifle ranges, and my department has a couple of contracts concerned with studying this. Impulsive noise in the environment is a special problem as compared with other sources of noise, and some work that we are undertaking is to look at the correction factors that are needed to estimate the level of disturbance from impulsive noise. Preliminary results indicate that about 15 per cent of all complaints made to environmental health officers in this country are concerned with impulsive noise and many of these are concerned with noise from rifle ranges. Regarding design criteria, it looks as if one should add a penalty of about 9 dB to the L_{eq} level in order to give you an equivalent level of annoyance.

1.20 H.R.N. Martin
Kingston Polytechnic

Dr Croome talked about different people varying in their degree of susceptibility to noise, and showed graphs converging at high volumes of sound. I should like to ask him whether similar variations occur for the same person at different times (is one, for example, more susceptible with a hangover than one was the night before?) and what form the graphs might take in this case.

1.21 Dr D.J. Croome

Dr Bryan exposed thirty-four subjects, reading material of their own choice for 45 min, to a random selection of street noise, aircraft noise and industrial noise at 55, 65, 75, 85 and 95 dB(A) sound levels. The data obtained from all the subjects were averaged, and graphs were drawn of the mean annoyance score for the five different sound levels (see Figure 1(a)). The data were presented graphically by plotting the square root of the noise rating against noise level as shown in Figure 1(b). Notice that the judgements for the three kinds of noise source were not significantly different but the rating between subjects of the same noise varied considerably. The least noise-sensitive subjects could miss rating some of the quieter noises if absorbed in reading, whereas the most noise-sensitive found that even the quietest sounds intruded and distracted them. As an example of the intersubject variation, see Figures 1(c) and 1(d). It can be seen that the largest differences between the high and low noise-sensitive subjects occur at quite moderate sound levels in the range 50–70 dB(A). As expected, there is a convergence at high sound levels. In attempting to establish a crude character profile of noise-sensitive people from evidence collected in field surveys, Moreira and Bryan suggest that individuals with high noise-susceptibility may be personality types that show a great interest in and have empathy with others, have a great awareness of their environment and are likely to be intelligent and creative. Their studies show that neuroticism is not consistently related to noise annoyance.

Figure 1

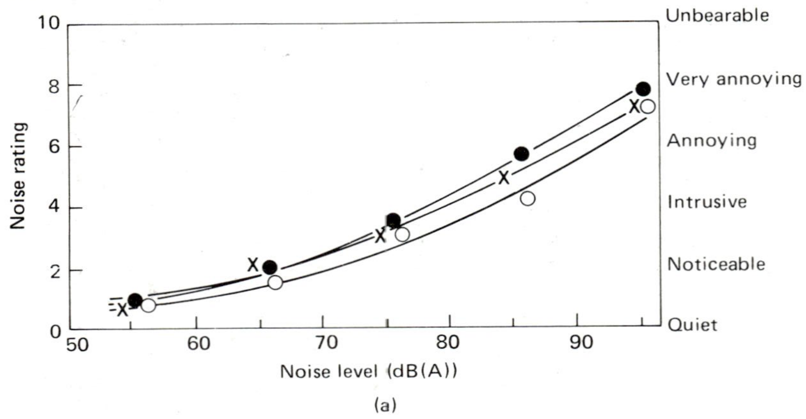

(a) Noise rating as function of noise level. Mean values for 34 subjects, for street (●–●), aircraft (○–○) and industrial noise (X–X)

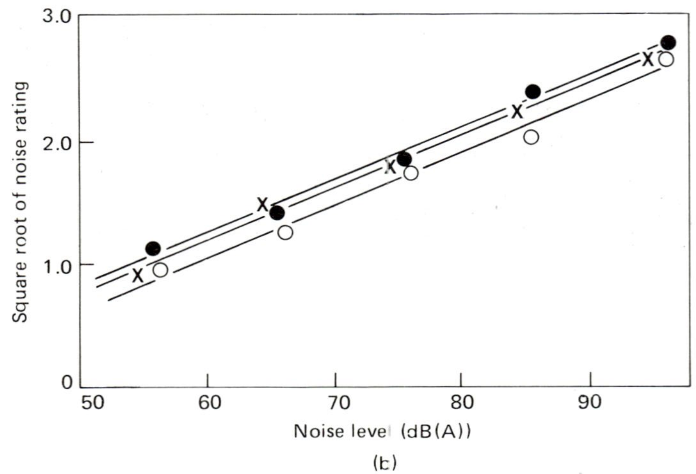

(b) Noise rating as function of noise level. Linearised mean values for 34 subjects for street (●–●), aircraft (○–○) and industrial noise (X–X)

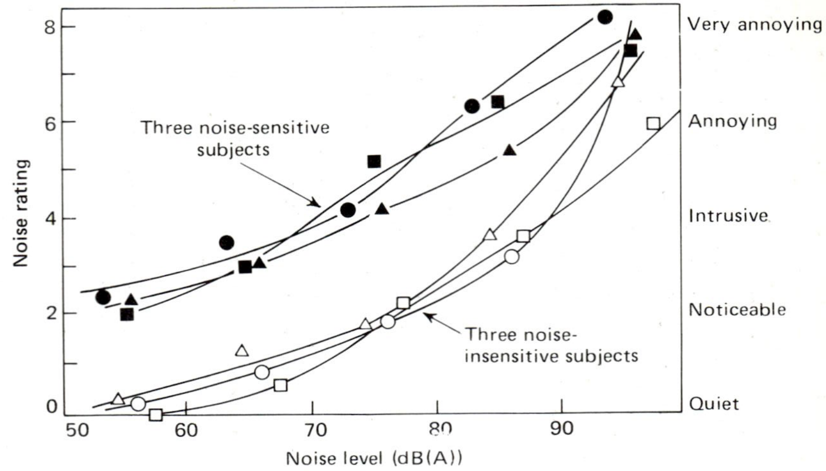

(c) Individual noise functions for 6 subjects (3 of the most noise-sensitive and 3 of the most insensitive to annoyance by noise): each curve is the mean rating of 3 noises

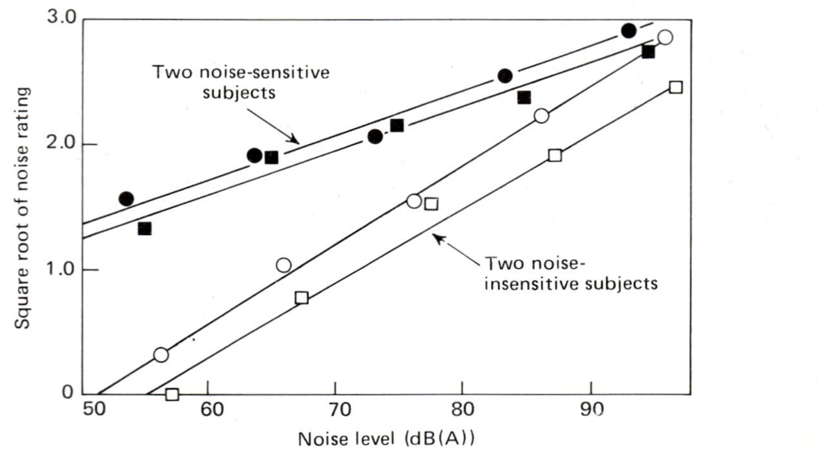

(d) Transformed noise functions for 4 subjects (2 noise-sensitive and 2 insensitive to annoyance by noise): each line is the mean rating for all 3 noises

1.22 Prof J.B. Large

I can confirm that at the high levels of noise there is a convergence between the groups that are relatively less annoyed (i.e., people in the middle range) and those who are very annoyed (i.e., the sensitive subjects). At low levels of noise, the psychological state of the individual at the time he is being asked to make a judgement of the noise has very much influence on his judgement, so you could get a relatively insensitive person who is in a sensitive mood, has a hangover or has had a row with his wife and he will respond differently. This is why I feel that the use of susceptibility in the setting of criteria is extremely difficult because, although you can quantify it statistically, you always have to keep in mind the great degree of variability between people, and for the same person at different times. It is only when you reach high levels of noise which start to produce physiological damage that the picture becomes much clearer, but at low levels of noise, where attitude, psychological state and a whole range of social, psychological, and economic factors come into play, there is enormous variability.

1.23 Dr D.J. Croome

In the chapter by Dr Bryan there is a mention of further work in which the subjects were allowed to carry out a variety of mental tasks of differing complexity, at their preferred listening level. There is a considerable variation in preferred levels. Some people work with music, some without, some in high noise levels and some in the quiet. People may prefer to work in the morning and others very late at night, which is, perhaps, an automatic regulator to help the human system to work at its most effective, tuned to the environmental pattern it prefers. The mathematician de Cauchey had to drink wine to achieve his most original work! Experiments are designed to take these variations into account by using many different subjects, for some of which tests are repeated at different times.

1.24 J.F. Bridges
Sound Research Laboratories

With regard to construction noise which requires measurements and predictions in terms of an L_{eq} level, I was wondering whether you had any guidance as to how you would determine a maximum L_{eq} level to set at, say, a site boundary in the daytime.

1.25 Prof J.B. Large

This is a very difficult question, mainly because it depends upon the construction site. If you are thinking in terms of a construction site in which piling is taking place and where you have a high degree of impulsive noise, then the remarks that I made to Mr Simpson about having to add a correction factor certainly must be taken into account. If one is thinking in terms of a global L_{eq} of 65–70 dB as being a reasonable daytime noise environment which should not be exceeded at the site of the construction boundary, then the L_{eq} level should be lowered if the process has a high degree of impulsive noise. How much it should be lowered, I do not really know at this point, but my guess, on aggregate, would be about 5 dB; in other words, if you set 65 dB as a normal workable environment and you have a lot of piling going on, and the piling has not been suppressed in any way, you should really aim at something like 60 dB. During the construction process the impulsive processes like piling are the most disturbing, from both noise and vibration points of view. The other aspect of construction noise which appears to be highly undesirable from a community angle is duration, but this is a matter of good housekeeping, management and consideration for others.

1.26 J.F. Bridges

You mention levels of 65 L_{eq}, which is a fairly generalised sort of level, and I wonder whether you should, perhaps, make some assessment on the pre-existing background, similar to the BS 4142 assessment. I was involved with a site in a very quiet area, where the steady L_{90} was about 35 dB(A): twice a day a train came through. The local council assessed an L_{eq} of about 60 dB(A) as acceptable for mining and clearance operations. This did not seem right to me, because the L_{90} background was very much lower than this.

1.27 Prof J.B. Large

I gave you my answer on the basis that we were talking about a site in an urban situation, where L_{eq} = 65–70 dB(A) is usually acceptable, but if you are talking about a rural environment, then certainly one should take into account the background noise, i.e., an L_{eq} that does not add significantly to the existing environment. The peak noise-level event should also be taken into account and here, I think, we are talking about a peak to background noise level of about 25–30 dB.

1.28 R. Taylor

Could I just add two points? The first is that we must not get confused between outside and inside sound levels. Professor Large is referring to sound levels outside and we assume a reduction of about 10–15 dB from outside to inside.

The other point is that I think the answer to the knotty construction noise question is a legal and not an acoustical one, because the law recognises that construction is a necessary operation which is unavoidably noisy and has to be carried on. There simply has to be a balance between the advantage of carrying on construction, provided it is done in a reasonable manner, and the effect on the enjoyment of the property of the person suffering the noise. The usual approach to it is to consider what is a tolerable level in all circumstances for the person suffering the noise, which usually means the level which enables him, while suffering a bit of discomfort because his neighbour has to do some construction and he therefore has to keep his windows closed, to communicate with other people in his office or on the telephone. In this respect, it does not have anything to do with the pre-existing background level and it is not really anything to do with the nuisance-assessment approach that has been discussed. It is simply the matter of two neighbours trying to live satisfactorily while one is carrying out a noisy operation which is useful to him only.

1.29 Prof J.B. Large

I do not want to discuss the legal situation, because the answer I gave was from an environmental and acoustic point of view, although what the law stipulates is right. Over the last few years we have been surveying rural noise situations and they are quite different from the urban noise situation. Curiously, you find that people who live in the countryside are more noise-tolerant than the people who live in towns, mainly because they do not have as much noise and therefore it is not so much of a burden but also because in many instances they find that noise is a method of social communication. If you hear the garbage collector or the postman, or the train in the distance, you know that there are people around and that there is life going on, i.e., it keeps you in touch with the outside world. Regarding construction noise, I think that one should aim to get satisfaction by designing your construction process so that it does not add significantly to the noise burden that already exists in the environment. So far we have talked about pile drivers, planes, trains and other everyday sources of noise. If you talk about noise environment, particularly in the countryside, a very important component is the natural noise of the wind in the trees, which most people do not hear but certainly contributes to the overall continuous equivalent sound level. When I am talking about the acoustic environment in the country-side, I am referring to the total sound environment, not just that from the truck going by, but also those sounds generated throughout the whole day by whatever process is occurring.

1.30 Dr A.F.C. Sherratt
Thames Polytechnic

Since 1974 and the Health and Safety at Work Act the noise in work places has come into sharp focus. In Chapter 1 Dr Bryan indicated that something in the order of a million people are employed in industries where noise levels are higher than 90 dB(A), which is the level the Factories Inspectorate accept for a normal working day. Discussion with the Inspectorate indicates they are keen to reduce the acceptable noise level they will allow for a working day. Of course, much of the problem is caused by the nature of the activities people engage in, in work places. I wonder what the possibilities are for reducing the allowable noise level. Is there a real possibility of reducing the high noise levels in work places or will more and more people have to wear ear defenders?

1.31 Dr D.J. Croome

The reader should refer to the treatises by Burns, *Noise and man* (1968), and Kryter, *The effects of noise on man* (1970), for detailed accounts of deafness and physiological hearing damage. Most people live and work in environments with noise levels less than 90 dB(A) – the starting point where damage risk is at present thought to be imminent. With regard to this level of 90 dB(A), brief mention should be made of work by Schwetz *et al.** which shows that hearing at a level of 75 dB(A) may take two hours or more to return to normal; the need for lowering this 90 dB(A) starting point should be considered. For most people the danger of exposure to high noise levels arises from traffic, and for the teenagers from discotheques as well. I think we have to put a lot of thought into trying to reduce noise levels and concentrate more on design so that noise control in vehicles and machines is an important feature of design. Prevention is better than cure.

* Schwetz, F. *Mschr Ohrenheilk. Lar. Rhinol.*, 4, 162–7 (1970).

1.32 Prof J.B. Large

Could I just widen your question a little bit, because I think you have raised a very important point? If one looks at the noise burden on individuals, whether they work in a boiler factory or are school children, you find that the total 24-hour noise exposure is very similar. In fact, people who work in noisy industries have only a few dB more per 24-hour exposure than school children, and school children have just a few more dB exposure than housewives. In other words, out total daily dosage of noise is not too dissimilar and is in the order of the low 70s to the high 70s each day. Now, as you will know if you follow the philosophy of the Environmental Protection Agency (EPA) in the United States, who base their long-term thinking in the legislation very much on protecting people from a hearing loss, they have hypothesised that you have to produce a 24-hour L_{eq} of about 55 dB(A) in order to exclude 95 per cent of the population from any sort of ill-consequences from environmental noise, and in environmental noise I include occupational noise. We have tended to express noise considerations in terms of people in factories, in terms of people in the environment, and we have not really given too much consideration to the people in their totality, and I am afraid I disagree with the EPA philosophy, because the burden of information which leads you to this supposition of 55 dB(A) is virtually zero. It is a series of hypotheses placed on top of one another that reach this conclusion, but it raises a very interesting point and it also goes back to the comments I was making in a previous answer: if you are to look at the effects of noise on people then you not only have to look at the effects of noise on people in their workplace, in their place at home, in their getting to work, you also have to look at the noise the birds make, the noise the wind makes in the leaves, and the noise that our clothes make as we walk along, because, although we do not hear it, and so exclude it automatically, it is part of our life style. Clothes themselves cause considerable noise, such as the rubbing of our coats and trousers, the squeaking of our shoes – and they all add up to a considerable amount of acoustic energy. We tend to talk very eruditely and philosophically about the effects of noise on people, but we are a long way from really understanding it or being able to assess the total noise burden that we are all exposed to in our daily lives. It is a much more complex problem than just looking at people who work in factories, or people who live in houses or offices.

1.33 Dr D.J. Croome

This is a very important aspect that I referred to previously in the discussion. We do not understand how noise affects the nervous system sufficiently well for us to advocate definitive sound levels. Most of our criteria are, in Professor Large's words, 'a series of hypotheses placed on top of one another'.

SESSION 2

2.1 Dr M. Barron
University of Cambridge

A simple question for Mr Edwards, concerning the Sports Centre in the West Midlands. You mentioned a temporary baffle built of plywood. Could you please explain what you mean by baffle?

2.2 S.C. Edwards
Hoare, Lea & Partners

It was purely a construction made from plywood about 25 mm thick: marine grade plywood, fixed on a timber frame. This produced a barrier between the noise source of the louvres and the cottages. In addition, there were two sides to the construction and this formed a tunnel design. The air had to pass out of one end, this being arranged at right angles to the cottages themselves. To sum up, the construction produced two advantages – a barrier, and also a change in the direction of the noise source.

I did not mention what the noise levels were. Originally measurements gave NR 40 at the site of the cottages, which, if perhaps you again knock off 5, was giving NR 35 in the bedrooms. This was obviously totally unacceptable. The local Environmental Health Officer who was involved, as he would be, said that if NR 25 was obtained, he did not think anyone would complain. This level was successfully achieved after the changes had been made.

2.3 J.E. Ledgerwood
Hann Tucker Associates

I should like to ask Dr Croome about his empirical formulae for predicting sound-pressure levels in rooms in terms of fan speeds. To what extent are these formulae applicable to high-speed fans such as the types commonly used in high-velocity systems, and do you know why they are inaccurate at high speeds?

2.4 Dr D.J. Croome
University of Bath

The results give reliable predictions (± 5dB) for centrifugal fan speeds below 900 rpm. Higher fan speeds give less reliable predictions, because the turbulence increases. Adjustment of louvres in ceiling grilles can have a significant effect on the sound level measured in a room and this effect will be more at higher fan speeds.

2.5 J.E. Ledgerwood

In relation to the generation of noise by air movement at low velocity, I see the implication that low-velocity systems can be noisy. Looking at your graphs, I think it may be a bit misleading in that the graphs have different ordinates, and that comparison ought to be by overall level rather than by the characteristic of the curve itself. The high-velocity curves show an overall level of over 70 dB(A), whereas the low-velocity ones show levels of 48 dB(A), so I think the implication that the low-velocity system was noisier is maybe misleading.

2.6 Dr D.J. Croome

With low velocity systems you cannot assume that regular attenuation occurs (most current design practice assumes you can make this assumption), because there can be some generation within a low velocity system, as the characteristics of the curves show. The difference between the sound level at high (1440 rpm) and low (480 rpm) speeds is 70 to 48 dB(A); incidentally, the latter figure was obtained with a system which should have achieved a maximum sound level of 36 dB(A) in the lecture.

2.7 J.E. Ledgerwood

Are you saying that the generation effects that you are getting with your system are over and above the attenuation, or are they separate from that?

If you have an infinitely straight duct and ignore any break-in noise, then the sound will decay. What is happening in a system which has a high proportion of fittings and is installed in tight spaces is that there is more turbulence and so, consequently, there is more noise. If fittings are very close together, say under about 7 m, then any attenuation can be neglected, because the turbulence patterns from each group of fittings interfere with one another. We are talking about the generation and the decay of acoustic energy, so the effects are interrelated. On the other hand, they are independent effects because the same fan can transmit more sound to a space when installed in one system than when in another system, depending on the transmission pathways.

2.8 Dr D.J. Croome

There have been many measurements on various individual elements in ventilation and air-conditioning systems and, to some extent, in water-flow systems. The problem arises when you come to put together a system comprising these elements, because you cannot add up the results for the individual elements. In other words, the sum of the whole is not the sum of the parts, but the methods in the ASHRAE Guide, which is similar to the CIBS Guide, assume that it is.

2.9 J. Simpson
Department of Finance, Belfast

My question is directed to Mr Edwards. In the Director's Headquarters in Yorkshire, he said 'The slow-speed motor position was changed to provide a further speed reduction to an ultra-low setting, only just workable, and the sound levels were accepted with measured values close to NR 30'. Did this ultra-low fan speed affect the heat output into the rooms?

2.10 S.C. Edwards

I did perhaps mention that, because these were cooling units, fortunately the heating coils were larger than necessary and this did not, in fact, affect the output. It does mean, however, that if extra-high cooling capacity is required in the summer, then the units may have to be turned to a higher speed. Most people seem to be of the opinion that if ever the temperature became very hot, occupants would put up with the noise. This is quite an interesting comment, as it does reflect the kind of thing that happens, say, in the Middle East or in any hot country – if the air-conditioning is noisy, people will accept the noise because the alternative is to become very uncomfortable and hot. One of the problems with reducing the fan speed is that one can reach a critical point when the switching-on function will not work. Thus, if the speed is set too low (below 250 rpm), there is insufficient torque generated to overcome initial friction and to start the motor and rotation of the impeller.

2.11 J.E. Ledgerwood

Could Mr Edwards tell me how he explains to his clients the extra cost involved in over-designing building acoustics?

2.12 S.C. Edwards

One hopes clients are enlightened. Certainly, they are very critical if one under-designs, and somewhere one has to make a compromise. It is found far easier to justify the fact that one has built-in margins and has a satisfactory system than to make excuses for the plant when its capacity is struggling under high load conditions.

2.13 J.E. Ledgerwood

I have found that many clients often accuse acoustic consultants of over-designing when they are not.

2.14 S.C. Edwards

I think, in the end, when it is the client who says 'Look, I am not prepared to pay for all this acoustic treatment', the consultant must point out that if whatever is necessary is omitted, 'Be it on your head'. I think, then, the consulting engineer or acoustics specialist is relatively protected, in the sense that he did make the correct recommendations in the first place. This is usually a good argument.

2.15 J.E. Ledgerwood

Extending that argument a bit further – if we are dealing with noise from outside, would you, in fact, be designing to the most susceptible or to the average complainant?

2.16 S.C. Edwards

I think one designs for the average, in the absence of complete test and measurement data. In fact, in both these cases this is what was done. There is no question that noise was not considered externally in either of the Sports Centre projects. In the first case identified, it was found during investigations that the urban noise levels as quoted by British Standards and as stated in the Wilson Report were both some way from the measured 23 dB(A) of the urban environment. One would not expect a noise level as low as this, particularly on an A category road which was supposed to be a trunk road. A sensible design decision was made at the time to the effect that if NR 35 was met on the site boundary, this would be sufficient. After the analysis, the client accepted the fact that some additional acoustic equipment was needed and approved the necessary additional expenditure to provide the boiler-house acoustic door. In conclusion, is it sensible to suggest that one should carry out a 24-hour noise survey for every project that one designs?

2.17 R. Taylor
Rupert Taylor & Partners

There is never a reason for designing to external noise criteria without carrying out a noise survey. You can place no reliance whatever on the Wilson Report or BS 4142. Considering that you talk about £7000 for the remedial treatment and that a 24-hour survey costs in the order of £100–£200, I think it is very, very risky to try to guess what the background noise is going to be.

2.18 Dr D.J. Croome

In my view, we should be designing for communities of people, not only complainants, and should recognise that there are various degrees of susceptibility to noise within that community. It is unlikely that everyone will be satisfied by any form of environmental design, but, I hope, we could aim to provide a suitable noise climate for at least 95 per cent of the community.

2.19 S.C. Edwards

I think one of the conclusions – and I partly agree with what has been said, and what we in our practice now do – for any project that is going to run for 24 hours a day is definitely to undertake an acoustics survey and then build the acoustic system to produce the necessary performance. I think that if the building is not running 24 hours a day, then we do not have this night-time problem.

2.20 J.E. Ledgerwood

Returning to the sound level at 200 m from the nearest houses, presumably somebody initially did a simple calculation to see whether there would be a noise problem?

2.21 S.C. Edwards

Yes, they did, and we were 10 dB(A) out. We never really did get to the bottom of this.

2.22 Dr H.C. Pinfold
University of Liverpool

I wonder if I might use the blackboard for a moment? I recently had a student who was designing a centre for pop music which seems to parallel the case we are talking about.

His hall was like this in section:

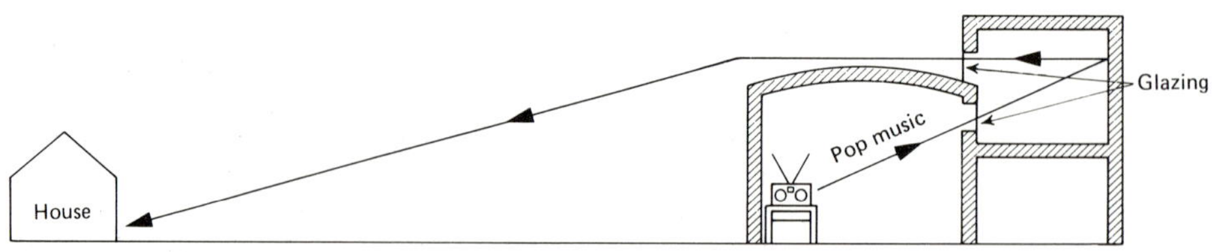

There was, as I remember it, a mezzanine gallery glazed to the hall and glazed to the outside above the hall. There was housing some distance away and he wished to know how to calculate the extent to which this would be affected by the noise. We never really solved this problem for, I suspect, the same reasons as apply in the case of the cottages. The prediction depends on starting with a known sound level at a certain distance and successively doubling the distance to get the decay.

2.23 R. Taylor

You are wrong to apply the inverse square law, since you must treat the source as a plane source. A convenient way of calculating the effect of distance on the sound level of a plane source is to use the formula.

$$SPL_r = SPL_0 + 10 \lg [(\sin \theta_1 - \sin \theta_2)(\sin \phi_1 - \sin \phi_2) - 6]$$

where θ_1 is the angle formed by a line drawn from the outer edge of the source to the receiver and from the receiver normal to the plane of the source, and θ_2 is the same for the inner edge. If θ_1 and θ_2 are in the horizontal plane, then ϕ_1 and ϕ_2 are their counterparts in the vertical plane. SPL_0 is the sound level as the distance, r, tends to zero. This formula gives zero dB drop until the distance is a/π, where a is the shorter side of a rectangular plane source – then 3 dB per doubling of distance up to b/π, where b is the longer dimension – and thereafter 6 dB per doubling.

2.24 Dr H.C. Pinfold

I would suggest that the source is rather more complicated than that. We start off with the loudspeaker. It is a point source but it creates a reverberant field. The sound comes through the glazing to the gallery (somewhat attenuated) and creates a new reverberant field. Reflections from this field go through the glass to the ouside (again attenuated) and are diffracted over the roof before falling towards the houses, attenuated further by the distance.

What I am saying is that this last attenuation by distance entails taking an initial distance of measurement from the source and this is more or less impossible to establish. I should suggest that you might be as much as 15 dB out.

2.25 Dr D.J. Croome

Another inaccuracy that arises in calculating sound propagated over distance is attenuation due to wind, rain, temperature inversion, and absorption due to ground surfaces. In practice, the 6 dB reduction per doubling of distance is more like 4 dB.

2.26 J.F. Bridges
Sound Research Laboratories

We have found that with pop music noise, if you are to design to an acceptable level, you must measure the pre-existing background in dB(A) and octave bands and try to design at least 10 dB below that, because the rhythmic noise from drums can cause much more annoyance than any dB(A) value will ever show. The environmental health people will get called in to a complaint and when you go along with a meter and just measure the dB(A), it does not really register as a problem, due to the effect of the 'A' scale filter on the lower frequencies.

2.27 Dr H.C. Pinfold

I agree. I have been involved in three court cases on that very problem and I agree entirely that you can stand outside a disco and not hear the music at all, but just the thud, thud, thud, which, as you say, does not show up on a sound-level meter.

2.28 J.F. Bridges

Generally, one can get noise problems in buildings from mechanical services, which are contractually difficult, because when you measure them you often find that they meet the design level which has been worked to, but, despite this, people are complaining about noise. This is particularly apparent at very low frequencies when you design to an NC or NR level, because the NC (or NR) figures start from the 63 Hz band. If you have got a rumble coming from some very large areas of ductwork, the problem will not show up on the NC or NR scale. The situation can therefore occur where the design level has been met and yet complaints are received of excessive low-frequency noise. I wonder whether the panel have had any experience of this problem?

2.29 R. Taylor

The PNC system does go down to 31.5 Hz, although people do not seem very keen on using it.

2.30 Dr D.J. Croome

The human ear is less sensitive at low frequencies ($<$500 Hz) and most sensitive at about 2000–3000 Hz, and the NC or NR curves reflect this. But the real point is that none of the available noise criteria covers every aspect of human response to noise. Complaints are not a very good indicator of a person's response to noise alone, because the complainant may have preconceived views about the given situation, which cloud the issue. An acoustic-social approach needs to be made to solving environmental problems. A further point is that the NC or NR curves do not take into account the character of the noise.

2.31 Dr R.K. Mackenzie
Heriot-Watt University

I should like to ask Mr Edwards a question on the hall design at the Centre for the Arts, Poole. You did mention in your chapter that lagging had been applied to the ducting and did I recall seeing a circular duct with cement lagging on it? I was just wondering if that is an area where there may be over-design. I am not necessarily saying that this was the case here, because I do not know the criteria you are working to, or exactly what noises there were, but circular ducts are incredibly stiff at the low frequencies and I think that would be an area where glass-fibre lagging might be adequate.

2.32 S.C. Edwards

There is duct insulation to prevent any regenerated noise break-out and this was constructed from dense 'stillite', with a Keene's cement finish.

2.33 Dr R.K. Mackenzie

Can I make one further observation? On Figure 6.7, you show a pipe supported by an anti-vibration spring hanger. I came across a case recently where such a hanger had been fitted, yet had been completely bypassed because of a regeneration of noise within the pipe, which occurred at a quite high frequency, around 500 Hz. Whereas the spring hanger was chosen in order to reduce low-frequency vibration from a pump, it was, in fact, regeneration from the pipe wall which caused the problem. It brings one back to the point that Mr Taylor was making with reference to coupled systems, one may have to consider secondary effects within the system.

Although some hangers incorporate neoprene pads in order to reduce high-frequency transmission, a gap in the frequency spectrum can occur, similar to a cross-over with loudspeakers, resulting in poor isolation at the 'cross-over' frequency.

2.34 P.R. Hartnell
Moniton Technic Ltd

I was also involved in the acoustic design for the Centre for the Arts, Poole.

Firstly, may I say that the amount of circular ducting used on the project was minimal. Although in most cases lagging circular ducting with mineral wool and Keene's cement is not necessary, as breakout noise from this type of ducting does not normally cause a problem, one has to remember that, in this instance, the noise criteria required in the space were particularly stringent.

Secondly, with reference to the isolation of vibration from pipework, the spring anti-vibration hangers used not only contained springs, but also incorporated neoprene in shear elements to control the audible range of vibration. In some cases, these types of hanger do not cover the noise and vibration isolation requirements. These awkward problems do, from time to time, occur and are normally dealt with individually, on their own merits.

2.35 R. Potter
SE Telecommunications

Mr Bridges of SRL emphasised the difficulties he has experienced with noise at low frequencies, a problem we have experienced with air-handling units in telephone apparatus areas.

Mr Edwards has told us of the optimistic claims made by manufacturers when designing for low noise levels, in the case mentioned NR 25. Our experience confirms his: indeed, we have great difficulty in achieving NR 45, let alone NR 25.

Could I, perhaps, bring Mr Redmore into the discussion and ask him if he has carried out any research on regulating and other types of valve as a source of noise, and ask Mr Edwards if he has used the velocities suggested without noise problems?

It has been our experience that these flow velocities, as applied to pipework, do produce noise, possibly due to velocity and pressure changes at regulating valves.

2.36 T.L. Redmore
University of Aston

I have not done any research work on regulating valves as, so far, my work has covered traffic noise, sound transmission through corridors and noise in ventilating systems. I would say, though, that there has not been a great deal of work done investigating noise in water systems, and there is a need for both more experimental and more theoretical study of the subject.

2.37 S.C. Edwards

Returning to the subject of telephone exchanges, I have some experience of those and have experienced problems with achieving NR 45. One of the reasons that this is often a problem, noticed when one commissions, is due to the highly reverberant nature of a telegraph exchange without its equipment. When the equipment is installed, the noise level reduces. When the plant is suddenly switched on, the concrete walls, hard ceilings, and ducts which are not insulated passing through the area add up to a highly reverberant space, and it has been found that at times there is a problem in achieving NR 45. This is another instance where it has paid to play safe and design to NR 40 to achieve NR 45. I can cite instances where one has designed for NR 45 and the result has measured NR 55. One of the main problems that we have found in telephone exchanges is regenerated noise from components producing high-pressure loss, such as automatic dampers, which, if not set correctly, can regenerate noise.

2.38 R. Potter

It is true that we have telephone exchanges with reverberation times of 7 sec or more, but allowance is made for reverberation in assessing the noise levels. Manufacturers are expected to achieve NR 45 with a reverberation time of 1 sec.

2.39 Dr D.J. Croome

Regarding flow velocities, I was always told not to design for more than 3 m/sec in water systems, whereas conclusions that can be made from current work are:

(a) The sound-pressure level is inversely related to the pipe diameter
(b) The shape of the fittings is more important than the number of fittings used
(c) The effect of a number of fittings is not cumulative
(d) Corroded pipes have been found to give higher sound emissions than clean ones
(e) Cavitation is established at a water velocity of about 5 to 6 m/sec, whereas in straight pipes it is not detected at water velocities of up to 14 m/sec
(f) Norwegian measurements suggest that higher water velocities (2–4 m/sec) would be acceptable in many situations

Higher velocities have the advantages of providing larger pressures for circulation, and smaller pipe sizes and also assist in clearing air pockets, although a compromise has to be met between these benefits and the higher pump power necessary. American field experience shows that water velocities of about 2.4 m/sec (8 ft/sec) can be used in commercial areas, but should be a maximum of 1.2 m/sec (4 ft/sec) in houses, flats and quiet areas of buildings to avoid noise problems.

2.40 R. Taylor

We have been talking about sound levels of NC 20 and NC 15. May I ask others to make a plea to instrument makers to make sound meters that measure as low as NC 15? It seems that, with the advent of ½-in microphones, sound meters seem to have a lower limit at about 20 dB. We need more people to ask the instrument makers for that problem to be taken into account.

2.41 D.A. Russell
Rosser and Russell

Just to make a comment about water velocity in the pipes in district heating schemes, it is important to provide some form of pressure reduction, either by controlling the main pumps or by using diverting valves, otherwise the pressure reduction required through the control valves becomes very high in low load conditions and the high velocity through the valve is bound to generate noise.

On the question of using increased velocity in pipe circuits without using extra energy for pumping, it is desirable to reduce sizes and increase the velocity in subsidiary circuits, while maintaining moderate velocities in the index circuit. This is an economic solution which does not use extra energy and makes regulation easier, since it is not necessary to lose so much pressure through regulating valves.

2.42 Dr D.J. Croome

One of the things that comes from the chapter by Mr Edwards, and this brings us back full circle to that by Dr Bryan, is the fact that there is an individual response to the environment. Some people will not even accept NR 25. One noise problem I encountered was on a suburban estate just outside Leicester, where a married couple lived in a semi-detached bungalow. They asked me to sit down and just listen to the couple next door, who had two or three children; but, besides children, they themselves led a completely different life style. They enjoyed a row or two each evening, which contrasted with the quiet neighbours. Noise was causing the quiet couple real stress and tension, to the extent that they were considering leaving their bungalow. Other people could have the same bother and it would not affect them at all. There is a limit to adaptation and how much noise people can get used to, which still causes a lot of stress and of which people may not be aware.

SESSION 3

3.1 Dr M. Barron
University of Cambridge

What is meant by field incidence, when we are talking about the mass law?

3.2 Dr R.K. Mackenzie
Heriot-Watt University

When referring to field incidence, one is talking about an integration over the angles at which the sound impinges on the wall under normal conditions, and this tends to range from 0° (i.e., normal incidence) to 78°, whereas 'mass law' assumes incident sound waves ranging from zero to 90°; so basically the field incidence mass law is simply stating that sound is not impinging at grazing incidence under normal field conditions, although it can under laboratory conditions.

3.3 Dr D.J. Croome
University of Bath

Over the mass law range the sound reduction index is given by:

$$R_\theta = 10\,\lg\left[1 + \left\{\frac{\omega m \cos\theta}{2\rho c}\right\}^2\right]$$

for a structure mass m, density ρ, angular frequency ω, for sound incident at angle θ. At normal incidence $\theta = 0°$.

$$\therefore\ R = 10\,\lg\left[1 + \left\{\frac{\omega m}{2\rho c}\right\}^2\right]$$

For random incidence:

$$R_r \triangleq R - 10\,\lg\,(0.23R)$$

In practice, sound is incident from 0 to 78°, hence:

$$R_r \triangleq R - 5.$$

3.4 P.H. Andrew
Kennedy & Donkin

Could I ask Mr Spring a question, and I think other people could possibly help with this? In many structures you bury, more or less, the supporting springs or rubber-type vibration isolators beneath massive blocks of concrete, which are virtually permanent. Has anyone any experience of ultimate failure of these members which are buried so far out of sight and possibly out of mind? Has any trouble been experienced with these structures?

How long do resilient mountings last?

3.5 N.F. Spring
Sandy Brown Associates

When the National Physical Laboratory first built a room floating on rubber isolators, they made provision for replacing the isolators in case they failed. No replacement was ever required. Designers do not now worry about the durability of rubber any more than they do about other materials used in building construction. Ozone degrades natural rubber, but this should not be a problem unless the isolator is exposed to sunlight or the ozone from electrical discharges. So far as steel springs are concerned, I have never heard of problems arising from rust.

Resilient mountings last about 50 years for natural rubber. Designers often build in fail-safe measures so that, should the resilient element fail, the building will not be damaged or unsafe. As far as I know, no resilient elements have failed.

3.6 Dr R.K. Mackenzie

I think designers must always be aware of the question of maintenance and thus the question of entry into machines, especially if you are mounting such items in pits, where materials may accidentally fall into the pit and even short-circuit the spring. Maintenance is very important. It is an area that is often underestimated, and materials such as acoustic ceiling tiles are often misused. There are so many theatres that have perforated tiles which have been painted that I have lost count of them, and it completely destroys the acoustics. I think we have a responsibility to give guidance to maintenance staff, because they obviously do not appreciate the acoustic aspects and this also applies to services.

3.7 P.H. Andrew

Where you have a pit one would expect that in some cases it might gather water, for instance, which also could short-circuit the isolating members. If you start making holes to drain the water, you are going to have further difficulties in maintaining the noise-attenuation properties.

3.8 N.F. Spring

Where this is expected to be a problem, one can obtain proprietary drains with flexible connectors, which will not degrade the insulation of such a structure.

3.9 A.H. Middleton
ISVR

A question to Mr Box, please, concerning Figure 9.6 in the chapter by Mr Simpson. It seems to me that there is potential for considerable confusion on this figure, because it seems to mix up a number of things. I gather that the curves for the laboratory test of the window relate to transmission loss, the site survey curves appear to relate to noise reduction and therefore are not necessarily directly comparable at all, and for C, the sound-attenuating air brick, we do not know what it relates to, whether it is a genuine transmission loss test or what. Can Mr Box give us any more information?

Could I make the point of how careful you need to be in interpreting a mass of figures in dBs that you might obtain for materials or a building and how you must make sure you know what you are actually reading?

3.10 R. Simpson
Yorke Rosenberg Mardall

Mr Middleton has correctly drawn attention to the fact that, in Figure 9.6, curves A and B (showing the laboratory test results) recorded the normalised level difference, whereas the site survey curves D and E relate to a direct transmission loss. While this confusion is unfortunate, it is felt that the conditions applying in this case are not such as to produce a seriously misleading comparison. In fact, we have since converted the laboratory results into direct transmission loss so that comparison can be made with the site results and, for this purpose, we have used the method described in BS 5821. We have found from this that the overall performance at the site is slightly better than the result obtained in the laboratory, as shown in curve B.

The graph for the performance of the attenuating air brick was taken from the manufacturer's publication and gives the transmission loss as a normalised level difference.

3.11 J.E. Ledgerwood
Hann Tucker Associates

Following on from the earlier contribution, I wonder if someone would like to comment on the use of mineral fibre as an isolating medium for suspended slabs.

Secondly, with reference to PNC curves, I think it ought to be pointed out that there is a cost implication in using these as against using NC or NR curves, because of the extra insulation or treatments required to meet the low-frequency requirements.

I was interested in Dr Mackenzie's comments on sound-pressure-level data. I would agree completely with those comments and I think that they may also apply to sound-power-level information, in that sound powers may have been converted from sound-pressure levels measured in that sort of fashion.

My last point is related to leakage of sound through small gaps in structures. There was an interesting graph shown by Dr Mackenzie, showing quite a large loss of insulation over the whole frequency range, through a relatively small gap. This seems to conflict with one's experience that small gaps tend to give greater loss at high frequency and less loss at low frequency. I wonder if he could explain that anomaly.

3.12 N.F. Spring

For domestic use mineral fibre is satisfactory, but for any stringent requirements at low frequencies I do not think it is satisfactory. The mass/spring resonance of mineral-wool floating floors is often about 80–100 Hz and, although it is well damped, if you want good low-frequency performance, you are sometimes worse off with a floating-floor construction of that kind. If you are going to use mineral wool, note the warnings issued by the Building Research Establishment, as only certain types are suitable.

I do not think there is much evidence of it deteriorating with time. You should use the right grade, normally the long-fibre types, certainly not the kind that has the fibres running at right angles to the face of the slab.

One should not use such heavy loadings that the material is compressed far beyond its linear deflection region.

3.13 **J.E. Ledgerwood**

Data from manufacturers such as rockwool would indicate that you can get as low as 8 Hz natural frequency, using materials of the right density and thickness, and heavy loadings.

3.14 **Dr R.K. Mackenzie**

Concerning the question of PNC curves as opposed to noise rating curves, my feelings are fairly strong on this issue, because I have come across so many systems with low-frequency rumbles! As far as cost-effectiveness is concerned, it comes back to a point made by Stephen Edwards as to whether one should over-design in order to play safe. However, this is not a case of over-designing. By using the PNC curve you are designing properly to take account of the low-frequency rumble. What you may be pointing out is that you may be over-designing at the higher frequencies by using the PNC curve, but if one uses the NR curve, one is likely to fail to make allowance for the low-frequency noise in the system. I come back to this point that the NR curves are based on equal loudness contours and there is no correlation between equal loudness and equal annoyance. At low frequencies there is far greater annoyance that the equal-loudness curves would suggest, and the PNC curve is more reliable if you are going to have a potentially noisy system at low frequencies. I spoke to Rupert Taylor about this and I think there is a general agreement among consultants that they tend to keep an eye on both sets of curves and at the end of the day they produce their own specific design. However, I think what we are talking about here are guidelines for building-services engineers to follow and what I am pointing out is that NR curves can lead to problems that PNC curves will avoid.

The other point concerning sound-power-level data I totally agree with and, yes, it does relate back to sound-power levels. There is a great deal of confusion here and I must confess that very often levels without a reference base are supplied and one has to go back to manufacturers to find out whether it is sound power or sound pressure, and of course it may be related back, as you have said, to sound power from the sound-pressure measurement.

3.15 **N.F. Spring**

The effect of gaps in a partition is sometimes puzzling. In the literature, the transmissibility of an air gap is often said to be independent of frequency or even greatest at low frequencies when, in fact, common experience shows that with a leaky partition the effect always seems to be greatest at high frequencies. It is fairly simply explained, I think, by the fact that most partitions have greatest sound insulation at high frequencies so, if there is a defect in the partition, it makes its effect felt where the partition is otherwise of high sound insulation. You do not notice any degradation where the sound insulation is already low.

3.16 **J.E. Ledgerwood**

I was relating to the graph that Dr Mackenzie showed, indicating an insulation loss throughout the frequency spectrum, not just at high frequencies.

3.17 **T.C. Farnfield**
J. Roger Preston & Partners

I have one question for Dr Mackenzie and one for Mr Spring. I should like to ask Dr Mackenzie whether he is advocating the use of PNC curves for all applications or does he have specific applications in mind and, if so, what are they?

I should like to ask Mr Spring whether he has any experinece at all with the use of polycarbonate glazing and, if so, how its acoustic performance compares with that of glazing of a similar thickness.

3.18 **Dr R.K. Mackenzie**

Certainly, in the case of systems involving centrifugal fans or boilers, I would use PNC curves. You are normally talking about areas where you are trying to achieve a really low ambient noise level. These are the cases where you have a critical design target and you have to be very, very careful what you achieve. A low-frequency rumble will be annoying. Obviously, one can get away with an element of rumble in many environments, but in particularly noise-critical environments, where it is going to be very annoying, I think one has to adopt PNC curves, particularly for boilers and centrifugal fans.

3.19 **T.C. Farnfield**

Would designing to PNC curves affect any masking effect from building services installations?

3.20 **Dr R.K. Mackenzie**

You are asking if it would give rise to a degree of masking? Well, here we are talking about quite low frequencies, round about 30–40 Hz, and certainly below the range of speech communication. It is the annoyance, the low-frequency rumble that goes on during an 8-hour day, which gives rise to complaints, not interference with speech.

3.21 **N.F. Spring**

No, I have no experience of polycarbonate glazing. I have never been asked to look into it. Perhaps you can tell me where and why it is proposed to be used.

3.22 **T.C. Farnfield**

It is being used for cells attached to a court.

3.23 N.F. Spring

Is this because it is cheaper or has it any other advantages? I should imagine it should be no worse than glass for the same mass per unit area but I cannot really say. Mr Inman, do you know anything about polycarbonate as a competitor to your product?

3.24 T.C. Farnfield

I just wondered if you had any idea of its acoustic performance compared with that of glass.

I am not sure about the comparison of price. I believe it may be of a similar order to that of sealed double-glazing units.

The application I have experience of is security-related. There are, for instance, areas in a court house which require facilities for visual surveillance, while retaining acoustic privacy.

3.25 C. Inman
Pilkington Bros

Polycarbonate has about the same density as glass and, therefore, by the mass law, has roughly the same acoustic performance. The reservations on it really are because of its flexing. If you push the centre of it with reasonable force, then it is possible to spring it from an aperture, like a window. Also, it is not very resistant to scratching. From a mere acoustics point of view, it performs the same as glass does.

I would like to comment on some points raised by Mr Spring and Dr Mackenzie. First of all, dealing with sulphur hexafluoride in double-glazing units, the sound-insulation performance of it is very much dependent upon the concentration of the SF6: a concentration of about 20–25 per cent gives the optimum performance. Another consideration is that the improvement is at mid-frequencies, that is, away from the low frequencies where traffic noise is important, and away from the high frequencies where strong components of aircraft noise are found. Therefore, it meets neither of these requirements. Furthermore, there is a strong resonance at about 200–400 Hz, which destroys the improvement that is obtained at mid-frequencies.

Laminated glass does damp out, to some extent, the coincidence resonances and we have found that interlayer thickness has only a second-order effect. Contrary to what is stated in some literature, it does not matter whether the interlayer thickness is 15, 30, 60 or 90 thou, the influence is only about ½ dB from the thinnest to the thickest. Multiple interlayers, however, do have a greater effect. In double glazing, the influence of laminated glass is about the same as if two different thicknesses of ordinary glass were used. That is, the use of 6 and 10 mm glasses would give about the same insulation as 6 and 6 mm laminated glass, and by laminating the second pane, that is, using two panes of laminated glass as Dr Mackenzie was suggesting, you have reached a saturation level and there is not really very much more work to do in damping out the residual resonance. The incremental gain is therefore very small.

As a last comment, Dr Mackenzie did say that, for sound insulation purposes, only one window would be recommended, namely, a wide air-spaced double window. In general, this is true, but I think it is a matter of 'horses for courses'. Sometimes it is possible to install, say, a double-glazing unit with a thick glass incorporated within it and get adequate performance, but there is a distinction between adequate and extremely good acoustic performance. Where there are other considerations, such as space requirements, or where the dominant requirement is thermal insulation, a sealed unit with one leaf of thick glass can give acceptable performance.

3.26 Dr R.K. Mackenzie

I have always felt that the cost of the thick glass made the use of it prohibitive, as compared to a secondary window. I take your point that every situation must be treated in isolation, particularly if you are talking about a development with many windows involved. I know it is certainly useful for traffic noise but it is possibly expensive.

3.27 Prof B.L. Clarkson
ISVR

May I ask Mr Box if he has done anything on ground vibration? Presumably the railway has all-welded rail but the road could give rise to some ground vibration.

3.28 K.E. Box
Yorke Rosenberg Mardall

No special precautions were considered necessary in respect of ground vibration. The railway, which might appear to present the worst problem in this respect, follows a sharp curve where it passes this site and trains run here at a slow speed, so that there is no significant ground vibration.

3.29 Dr M. Barron

I might as well say what I know, which is that ground vibrations were not taken into account. In terms of this particular site, I would feel that the major and only serious source for ground vibration might be the trucks going to the building depot. The trains travel extremely slowly on these railway lines, certainly the one that passes adjacent to the site and, as Mr Box mentioned, the problem is basically wheel squeal as they were going round the corner at very slow speeds – about 5–10 mph. Certainly, the problem which would worry me most about this site is these trucks, because I, personally, would not want to be woken up at 6.30 in the morning by trucks thundering past, and it is one of the cases where there is no money available to compensate for nuisance, because it does not come up on an 18-hour survey.

3.30 Prof B.L. Clarkson

I raised it because in some of the recent TRRL traffic-noise surveys buildings close to heavy traffic experience very low-frequency vibration from airborne sound, caused by the exhaust systems of the vehicles.

3.31 Dr D.J. Croome

Can you describe how you treated the internal noise problems in the dwellings, including a reference to the heating system?

3.32 K.E. Box

In planning the dwellings (Grafton 3A Housing, Kentish Town, Highgate Road, London NW1) rooms of like function were arranged on either side of party walls, so that, for example, sleeping in one flat would not be disturbed by the neighbour's TV.

The party walls between dwellings are loadbearing, dense concrete block, 190 mm thick, plastered on both sides.

The party floors between maisonettes in the four-storey buildings comprise 150 mm reinforced concrete slabs (for fire resistance) finished on top with 'Durabella Conti wood'. This finish consists of 20 mm tongued-and-grooved chipboard fixed onto softwood battens. The battens are spaced at 600 mm intervals and are lined underneath with polyurethane-foam pads. The manufacturer claims a Class 1 impact-sound reduction when the material is laid on a concrete slab 115 mm thick.

Ove Arup and Partners prepared the following written reply concerning services:

During the preliminary design stage, due consideration was given to the control of noise and vibrations from the mechanical and electrical services. Advice given to the architect on these aspects covered the location of the central boiler house and the services distribution system throughout the site. It also included the distribution of services within the dwellings.

The following are some of the steps taken to arrive at an optimum solution with respect to the control of noise and vibration:

1. The location of the boiler house not only satisfied the engineering requirements of centralisation but was also remote from occupied areas. A retaining wall constitutes one side of the boiler house.
2. Atmospheric-type boilers were selected for their simple operational and maintenance characteristics, and also to avoid noise nuisance during firing.
3. Within the boiler house, circulating pumps are supported on inertia blocks and fitted with flexible inlet and outlet connections. The pump motor is also fitted with flexible connections.
4. Heating distribution mains are either buried or located in a services void, outside the dwellings.

Where necessary, pipes are supported on vibration isolation hangers.

3.33 A.C. Skelton
The Post Office

Mr Box outlined the extensive noise tests carried out on the site. Did you do anything about atmospheric conditions in that it is hemmed in by railway, road and also the building site? Following from that, you made particular reference to the fact that the hardwood surrounds to your windows were stained, so that you could keep the good neoprene seal. Have you made certain that that is going to be carried on at maintenance stage?

3.34 K.E. Box

No special precautions were taken in respect of atmospheric pollution. There were no grounds for supposing that this site was significantly worse than, say, any other site in Camden. The one serious problem was noise pollution.

At the end of the contract, a maintenance manual was prepared for the client. The manual contains advice on the upkeep of the buildings and includes the stain manufacturer's instructions for aftercare. It is unlikely that the client would change to a painted finish, which would be more costly.

3.35 A.C. Skelton

Mr Spring made mention of using speakers to cancel out loud diesel exhausts. What sort of noise would come from the speaker to cancel these out?

3.36 N.F. Spring

It would be a wave form that was complementary to the sound being generated. This is one of the problems, of course. You need a loudspeaker that will radiate efficiently at low frequencies and you need a highly-rated power amplifier to effectively absorb the sound. These are the main requirements.

3.37 Prof B.L. Clarkson

Perhaps I could sound a voice of warning here. Active noise control relies on radiating the equivalent volume of sound to cancel out the unwanted sound at a specified point, but in some parts of the field you might get double the volume. This is generally a rather difficult development problem.

3.38 N.F. Spring

Generating 'anti-sound' is just one way of describing the situation. Another way of looking at it is to say that you are reducing very much the impedance near the source, so that it is not capable of radiating the sound which it otherwise would do, because it has got nothing to work against. If you take that view, then it is perhaps easier to accept that the system does not necessarily have to transfer the sound energy elsewhere, and no energy conservation laws are violated.

The demonstration by Professor Chaplin (of the University of Essex) was very convincing but it is a pity that he has not yet published anything on his techniques. Until he does so, I think there is bound to be a certain amount of scepticism.

3.39 Dr R.M. Mackenzie

My personal opinion on anti-noise, phase cancellation, is, I think, that we are perhaps in great danger of overselling this method. I think quantity surveyors will probably tell us, once somebody starts using it for a practical application, that it is just a non-starter for most building elements. To begin with, it works better at the low-frequency end of the spectrum dealing with discrete modes. It has suitable applications in defence work, for example in naval architecture, where it may be applied to cancelling out resonances from submarines. Certainly it has been used in the aircraft industry. I understand Southampton have been involved in applying it to the cockpit communications (unsuccessfully). It has applications in the case of very large expenditure buildings, for example nuclear reactors, but I think it is premature to talk about using it in ventilation ducts. You have to remember the cost of amplifiers and vibrators, and to start using that in place of lagging to a duct is really absolute nonsense, in my opinion. I think it has severe limitations, certainly for the domestic-walls situation, where I see it as an absolute non-starter, considering the alternatives to increasing the insulation, but it has definite applications in particular environments for particular machines.

3.40 A.H. Middleton

We have been developing anti-noise systems at ISVR, mainly, as Professor Clarkson said, in connection with an application where the noise is heard through ear muffs. That is, in aircraft where the pilots have difficulty in hearing the intercom, for instance. You can practically radiate a cancelling signal into a cavity like an ear muff and get a very good effect. You can get 15 dB(A) or more effective noise reduction over the frequency range which is a problem inside aircraft. That works very well. You can also do it in ducts for generator sets and things of this sort, and get a very reasonable attenuation. Quite a lot of people have done this and shown that it is practicable. It gets more difficult as you get more flow going down the duct, because you get convection problems, but I agree with Dr Mackenzie that it is going to be entirely a non-starter from the point of view of cost and complication in any normal building application. One other thing on the anti-noise side, that is slightly verging on the same subject, is the business of 'noise perfuming', where you add a bit of noise to make the situation more acceptable, to avoid disturbance by some other intruding noise that you might otherwise have heard. This can be useful on occasions.

3.41 H.R.N. Martin
Kingston Polytechnic

I would like to ask Mr Spring a question on degration. He said earlier that the addition of polystyrene lining or the use of permanent wood-wool shuttering can lead to a degradation in sound insulation of the main structure. How can you predict the amount of this degradation and the frequencies at which it takes place?

3.42 N.F. Spring

You can estimate the frequency at which the degradation will occur just be considering the system as a mass-spring resonating system, with the outer plaster layer as the mass and the lining as the spring. I do not think it is possible to estimate in advance the extent of the degradation. One has to rely on case histories. This effect was not predicted in advance of its occurrence. It was observed and then acoustic theory was applied to provide a plausible explanation. It is not often in building acoustics that theory predicts the occurrence of new phenomena.

3.43 Dr M. Barron

A question to Dr Mackenzie – you mentioned that a masonry wall does not conform with the mass law, whereas, whenever I have seen graphs with the line of the mass law and blobs on it for different constructions, I have always noticed that brick walls and so on seem to fall very close to the mass-law prediction. I wonder if you would comment.

3.44 Dr R.K. Mackenzie

They run parallel to the mass-law prediction, yet they do not agree with it. It is by a partial coincidence that it does actually increase at that rate. The mass, of course, is one of the terms in calculating radiation efficiency, which is one of the important factors in calculating the true transmission loss, but you will notice that in Figure 8.6b the curve which shows the field of incidence mass law is parallel with the true transmission loss curve and this curve does allow for the mass impedance, among other things. It is the other things which reduce it to below the mass law, so it is quite true that the curve does increase by a factor of 4 to 6 dB per doubling of mass, but it does not agree in level with the mass law. The mass law takes it a good 10 dB higher. The wall, of course, is resonance controlled and one should also bear in mind that the coincidence in many of these cases occurs at between 200 and 300 Hz.

3.45 Dr H.C. Pinfold
University of Liverpool

Anybody wishing to get some experience for themselves of anti-noise can go home and misconnect one of his sterio loudspeakers. He will find that the bass resonance does not disappear, but it is considerably attenuated when you connect the two loudspeakers out of phase.

3.46 Dr R.K. Mackenzie

Could I just say that this applies only if the frequency response of your two loudspeakers is identical?

3.47 **Dr H.C. Pinfold**

They are usually fairly closely matched at low frequencies.

3.48 **W.T. Gracey**
Gracey & Associates

We have had dire warnings over the last two days about sound insulation and not relying on manufacturers' figures. Can anybody comment on floating floors? Do they normally meet manufacturers' figures?

3.49 **N.F. Spring**

Floating floors sometimes work well. Sometimes they do not. Generally, the cases brought to my attention are cases of failure, because people do not need advice otherwise, so I probably have a distorted view. As is probably now well known, the BRE have carried out surveys on dwellings which show that a high proportion fail to meet the standard. The results were largely attributed to an unsuitable resilient layer or poor workmanship. It is difficult to see any reason why the situation should improve for new houses.

With regard to purpose-built floors in studios, etc., where the client can afford it, it is fairly easy to design out the problem. We have gone back and checked them. Sometimes they have failed because the client has decided that he cannot afford to install measures which would give, say, a 95 per cent chance of success.

3.50 **Dr R.K. Mackenzie**

I think that until some impetus is given to following up tests on floors we are not going to improve the situation. I mean, we are talking about a situation of 60 per cent of all floors in dwellings failing to comply with the regulations. The last time I tested a floor and it passed the Building Regulations was a good three or four years ago and I test around thirty floors a year, so the situation is very, very bad, but there is no incentive to the designer or the contractor to improve his particular design or construction, because there is no check-up afterwards, and until we have a situation where there is going to be a check the situation will not improve. I am quite sure of that.

3.51 **Prof B.L. Clarkson**

While you are thinking of another question, can I ask Mr Spring to follow up on the floor? He showed a picture of a studio floor being constructed and I think he said that on top of the floating concrete there was some bitumen and on top of the bitumen a thin layer of screed. It seems to me that you are wasting your time putting bitumen in, because that certainly would not do any damping in that configuration. Bitumen will only dissipate energy in shear and should therefore be used in a sandwich construction. Adding damping will only iron out some of the peaks in the transmission, the average will not be changed very much, so when you look at a frequency average you probably would not see much effect.

3.52 **N.F. Spring**

That is correct. First of all, the damping layer is a bituminous compound. The optimum place to put it is in the middle so that it is sheared by the two layers of concrete, but again the structural engineer complains that the floor has lost structural strength. We calculated that in the situation I described with 150 mm basic floating floor, for 2 mm of bituminous compound and 50 mm of concrete screed on top, the damping would still be enough for our purposes. The main purpose is to avoid having a floor which vibrates too much when walked on. Otherwise there is a risk that floor-standing microphones of extended low-frequency response will overload the amplifiers. The incorporation of this damping layer seems to work satisfactorily.

3.53 **Prof B.L. Clarkson**

Dr Mackenzie talked about machines and wanted to know the acceleration at the mounting point of the machine. Now, will that depend on what you have mounted it on? So the manufacturers' data are no good to you, because the manufacturers do not know where you are putting it.

3.54 **Dr R.K. Mackenzie**

I think that under most circumstances the floor on which it is located is going to be of a fairly similar type. It will be some sort of concrete slab, and if manufacturers were to carry out tests on this type of floor (obviously, in exceptional cases the information would not be any use), I think some indication of the force at the feet would be better than no indication. I really cannot overemphasise the magnitude of safety factors that are built in simply as a result of having inadequate vibration information. You have no guide at all and it is impossible to predict the noise level underneath the floor except by experience, and the pity of this is that the theory does exist. If one has an idea of the approximate force coming into the floor, one can predict with a reasonable degree of accuracy the noise level underneath. Unless you have some indication of the forces coming down, you are not going to get anywhere, but I certainly take your point. The loading system is important.

3.55 **Prof B.L. Clarkson**

Perhaps I could just add that it may be true that if you are mounting a machine on a concrete floor, what you need from the manufacturer can be supplied, but where occassionally you are going to mount this machine in the roof space you may well not have the same kind of floor construction. It may be only some girders or some relatively lightweight structure just to hold that air-conditioning fan and that is, I think, when you are likely to get into trouble.

3.56 **A.H. Middleton**

I think this opens up a big area of discussion as to how you test a machine for its foot vibration output, if you like to call it that, but you say, Dr Mackenzie, you can do a reasonable calculation if you know the forces going in. Are you going to ask people to tell you what the moments going in are and all other forces as well, or are you just going to rely on vertical force inputs in your calculation? If so, do you think that is fair?

3.57 **Dr R.K. Mackenzie**

I think that if we know what the vertical forces from the feet of the machine are, we are certainly going to be within about 5 dB, at any frequency in the 63–4000 Hz range, of predicting what the noise level will be. The problem is not so much in resolving the level of vibration within the slab from a level of vertical vibrations entering. It is in knowing what is happening at the ends of the concrete structure. This is what has foxed most people until now, with the coupling losses at the end of the concrete floor. Various questions have to be answered. Is the concrete floor floating for a start? If it is not, how is it jointed onto the rest of the structure? Is it strongly or weakly coupled? How much energy is going elsewhere and how much energy is radiating underneath? These are the parameters that until recent times have been unknown but are now fairly well documented. Considerable information has been built up of coupling loss factors and, now that we have this information, the tools are there, and what we need is more information concerning the energy input, but I do not underestimate the problems which you and Professor Clarkson have alluded to. But, in the same way that a library of information has been built up on joints, if you consider all the types of joint in buildings, it means a fairly big exercise, but it can be done. In the same spirit of things, manufacturers could also build up information on specific types of suspension. There is not an infinite number of different suspensions for fans. They go on either a beam or a slab floor, or something similar to that, and I think that this information at the end of the day is going to save the client a lot of money.

3.58 **J.E. Ledgerwood**

A couple of points on this vibration issue. Firstly, I think that if we are in a situation where we have plant mounted over an area where there is likely to be a noise problem, it is most likely that that plant would be isolated, in which case information from manufacturers might not represent the sort of forces which would actually be going into the structure. The second point is that if force information were to be available and not response information, it would represent, presumably, a machine in pristine condition as opposed to one which may have been running for a year and would have bearing wear and so on, which would affect its vibration characteristics. I wonder whether anyone would like to comment on these points.

3.59 **Dr R.K. Mackenzie**

I agree with what you are saying and I have simply identified it as a problem. It is very difficult to know what to say on the second point of maintenance, because it does lead to a very big problem. Machines do go out of balance, and many things happen to them in fairly short periods as well. I think one can only recognise, with experience, potential problems with fans and boilers, and design the system to make allowance for these problems; perhaps that is where a consultant really earns his money. He knows the faults that can develop.

3.60 **Prof B.L. Clarkson**

To optimise the design you need to know the characteristics of the floor, the isolator and the machine, and you cannot pull any of these three independently out of a manufacturer's book.

3.61 **Dr R.K. Mackenzie**

In some respects, we are talking about a situation somewhat analogous to the sound-power/sound-pressure measurement system which is made in completely different environments, and one has to be aware of this environment. It is not unlike the case of vibration, but one would have to look carefully into the situation where the vibration measurement was made. What I am saying now is that we have nothing, we just have a total absence of vibration levels. At least with sound power/sound pressure we have a guideline and it is something to go on, but with vibration levels we have a situation where we have nothing to go on, and obviously, in the absence of information, we have to overdesign.

3.62 **J.E. Ledgerwood**

A number of manufacturers, particularly of refrigeration equipment and air-handling units, when they commission their equipment, do a vibration test, which is normally a measurement of the amplitude of vibration in an installed condition, which may or may not include anti-vibration mountings. This is, presumably, a check on whether the equipment is out of balance, and it seems to me that this sort of test is completely meaningless, because no account is taken of the way the machine is installed. I do not know whether anybody knows about any sort of standard which might be or could be drafted to improve this situation.

3.63 **A.H. Middleton**

I am afraid I do not know of any standards. I have given a lot of thought to the ways in which one might do it and it seems to me to be extremely difficult, one of the difficulties being that the vibration of the feet of the machine depends a lot on how you mount the machine. When you start mounting it in different ways and with different sorts of construction, you add extra constraints to the foot motion. One thing that I wanted to remark upon was the vibration isolation of things such as air-handling units and so on in buildings. I find a very depressing situation with regard to the knowledge, capability and ability of the suppliers of many anti-vibration mountings that are purported to be suitable for such applications – there is a very distressing degree of misapplication and misunderstanding of what mountings are supposed to do and what sort of mounting is suitable. An example is a total failure to take account of the longitudinal thrust of air-handling units with flexible connectors in ducts, so that the mountings fall over or go on to their end stops and hence do not isolate at all. Another one is the provision of mountings with springs which are inherently unstable, where, if you deflect them within their allowable deflection range sideways, they fall over and do not come back again. The third one is the effect of the thrust from not just the main air supply but also from other machinery: thrusts resulting from misinstallation of piping, misapplication of bellows in pipework and all sorts of similar side effects which make what might be good isolators completely useless.

3.64 G.A. Forbes
Stuart Forbes Ltd

The following precautions are important:

1. Reduction of vibratory noise transferred to building structures by means of mechanical support to piping, with particular reference to passage through walls/floors/ceilings, should satisfy the requirements of the 1976 Building Regulations that such perforations shall be 'fire stopped'. An insulating sleeve which satisfies BS 476, Part 8 (1, 2 or 4 hour fire delay) can be used as a fire stop.
2. Reduce vibratory noise transferred from air-duct supports to the building fabric.
3. Avoid creation of vibratory noise at support points to reduce noise within air ducts being accentuated.
4. Treat air ducts externally to avoid break-in and break-out noise.

SESSION 4

4.1 N. Jansen
British Rail

I should like to ask Mr Middleton, bearing in mind the development of sandwich-type roof and wall structures for factories and lightweight clad buildings, if there have been a lot more noise problems created by using these as opposed to asbestos cement, or similar, which was used in the past, particularly in relation to items such as roof fans and roof-mounted plant.

4.2 A.H. Middleton
ISVR

No, I do not think that the roof fans and roof-mounted items are any more of a problem in that sort of structure than they used to be. It is usually direct sound radiated from the fan itself that is the problem, rather than any interaction with the roof structure. The one thing that may sometimes cause problems is that the transmission loss for sound through the walls may not be as high as for the old-fashioned heavy brick-built type of factory.

4.3 B. Day
University of Bristol

I have met a problem concerning the handling of constructional steel work. This was in a new factory, which was constructed from plasterboard-faced insulating panel on framing. There were two sheds alongside each other, one new and one old. We measured a reverberation time of about 6 sec in the new shed and about 1½ sec in the old one, where the brickwork was covered in dust that had accumulated over the years and which provided enough absorption to keep the reverberation time down and actually seemed to help solve the noise problem. I disagree with Mr Middleton on this question of the value of absorption in factory spaces. Certainly, in that particular case, adding absorption did provide very great improvement, although it may have been a rather special sort of process.

4.4 A.H. Middleton

I was not trying to imply that adding absorption in factory spaces is never any good. My main point was that it is some good only where you are worried about a reverberant-noise problem, which in this instance appears to be the case. My point about the lack of value of absorption in factory spaces applies when the problem is a direct noise problem where the reverberant field is not significant.

4.5 R. Andrews
Andrews, Downie & Kelly

May I raise a question about hard surfaces generally, in all sorts of buildings? All our new buildings, our panelled buildings, seem to produce hard, straight surfaces. We have not talked about churches as auditoria, which I though might have been interesting – organs, people singing. Could you say anything about the development of hard surfaces in buildings as producing side effects in terms of noise that give us problems?

4.6 B. Day

It is interesting that you should raise that. We have a particular problem with Gloucester Cathedral Chapter House, where the Dean and Chapter would quite like to reduce the reverberation time and use it as an auditorium. At the moment the reverberation time is about 5 sec. What makes the problem difficult is that one just cannot fix absorptive materials onto the walls, because it is all stone and you cannot put Rawlplugs into Norman stonework. There is often a problem in hard-surfaced buildings. However, I have to confess that I do not understand why old churches often have reverberation times as short as they do. Tewkesbury Abbey, for example, has about 2½ or 3 sec and there is nothing, except some rush matting on the floor, that could possibly be providing any absorption. It must have something to do with the convolution of the surfaces, such as the detail of the surface moulding and the ribs on pillars. In modern buildings with similar materials the surfaces are often absolutely flat and the total surface area must be much less, giving very long reverberation times.

4.7 R. Andrews

Would anyone extend the discussion on factory noise, apropos materials? One knows that with flat roofs supported on a wood-wool decking, for example, the wood-wool itself acts as an absorbent because it is highly tessellated sound-absorbing material and sometimes produces better results than plastic-faced board or sandwich asbestos decking. Has anybody any experience of that?

4.8 A.H. Middleton

I have noticed that it seems much less usual to get nice listening conditions for orchestral music in a hall which has a proscenium and where the orchestra is, as it were, on the stage of a theatre. Do you think this is a general feature?

4.9 Dr M. Barron
University of Cambridge

Yes, there is no doubt that in a theatre you are presented with several problems when you want to use it for music. The most obvious one is that the reverberation time is too short, and the first thing that one should do is to include some form of enclosure so that the sound does not get lost in the fly tower, but is projected forwards. However, if you look at what your stage enclosure can do, you discover that really you do not want to accentuate the early sound because for speech your design will normally be organised so that there is plenty of early sound. What you really want to do is to provide more reverberant energy. However, it is very difficult to build something that will reflect the sound and for this sound not to land straight on the audience. So it is far from easy to provide anything really satisfactory, when you are starting off with a proscenium shape.

4.10 Dr H.C. Pinfold
University of Liverpool

May I suggest to those doing research on concert halls that there are two matters that might be worth looking into?

My recent area of interest is loudspeaker design and in this field it has been discovered (contrary to what is said in the textbooks) that we are able to discriminate amplitude variations in frequency response of as little as 1/3 dB.

Textbooks are very vague in their recommendations as to how reverberation time should vary with frequency. At one time they recommended that the reverberation time should be constant with frequency but later it was conceded that it should rise in the bass.* Perhaps we might find that there is a correlation between frequency response and the favour accorded to particular concert halls. Certainly, the world of hi-fi takes frequency balance very seriously.

The study of loudspeaker performance suggests another area of acoustic taste that might concern us. Some music has an imporant spatial content. That is to say, where there is a heavy polyphonic content such that groups of instruments carry on a dialogue with each other from different points in space, the ability of the listener to identify spatial relationships is important. This is less so in the case of large-scale Romantic music, where blend is all-important.*

A large reflector placed over an orchestral stage tends to kill directional information, as can a centrally placed loudspeaker in a stereo system.

* See Croome, D.J., *Noise, buildings and people*, Chapter 10, 530–3, for a review of recent theories on these matters.

4.11 Dr M. Barron

You raise a large number of issues. I do not know how common they are nowadays, but on a piece of Japanese hi-fi equipment there was a little control which called itself 'Presence' and in fact what it was giving was nothing more than a 5 dB boost or cut at about 1000 Hz. Listening to this in a hi-fi system made an amazing difference to the sensation of the sound. I do not know the answer, but on the other hand I feel very suspicious about the relevance of this to the actual situation of listening to music in a concert hall, because there the situation varies very considerably between one seat and another, and people do not complain. All musical instruments are directional and strings, for instance, radiate most of their energy at right angles to the sounding board, so, depending on where you are sitting in the hall, you will be receiving more or less of this, and this is not something that people generally comment on. To come to your question about the reverberation-time characteristic relative to frequency, somebody who supports your view about the sensitivity to this is Jordan and you can read about it in his book that has recently appeared, but I am rather uncertain about it: one never knows whether people are right in the end. I would like to say that the problem is much more complicated than you suggest, because frequency is a component in all acoustic measurements that we make, so we not only have the reverberation-time characteristic, we also have the steady-state characteristic, i.e., how loud the sound is at a particular seat, as a function of frequency. But we also have, if we start breaking it down into early and late sound, the question of how important the early sound is in the low frequencies, in the mid-frequencies and so on, and the problem is therefore an extremely difficult one to handle. There is no doubt that a lot more work can be done.

4.12 Dr H.C. Pinfold

I do not think I made my point very clearly, which is this. If you look at concert programmes over a period of time, you will see certain tendencies of musical taste and these change from time to time. I would suggest that these are paralleled by changes in acoustic taste, with particular reference to frequency response of favoured halls.

4.13 Dr M. Barron

My point was that whereas you use the expression 'frequency characteristics', in my field this can refer to a large number of different things. I am just saying that the problem is much more complicated than it might at first seem. For instance, quite a few studies have now been carried out in which recordings were made in actual halls and subjects were asked which recordings they preferred and their preferences then correlated with the physical acoustic characteristics of the halls. There have been two principal studies in Germany*, in one of which frequency did not prove to be important, and in the other the frequency characteristic of the reverberation time came out as the third most important characteristic.

* Lehmann, P., and Wilkens, H., 'Zusammenlang subjektiver Beurteilungen von Konzertsälen mit raumakustischen Kriterien', *Acustica*, **45**, 256–68 (1980).
Schroeder, M.R., Gottlob, D., and Siebrasse, K.F., 'Comparative study of European concert halls: correlation of subjective preference with geometric and acoustic parameters', *Journal of the Acoustical Society of America*, **56**, 1195–201 (1974).

4.14 R. Andrews

May I just ask either of you a supplementary question? What is the optimum for a concert? Is it a violin solo in the middle of a Welsh valley on its own on a Sunday morning? Where do you strike a balance for optimum satisfaction or optimum reception?

4.15 Dr M. Barron

The question is: 'How do you establish what is the optimum?' and there is no individual person you can ask. Beranek made a study, published in 1962* – he went round asking conductors and musicians. They do, of course, not sit in a typical seat in a concert hall. He then asked music critics, who are always put within the first few rows from the stage, so perhaps they are not good judges either. It is extremely difficult and about the only think you can go on is the general reputation of the hall over a long period of time and really, when it comes down to it, the person who decides is the acoustic consultant who is given the job.

* Beranek, L.L., *Music, acoustics and architecture*, John Wiley & Sons, New York (1962).

4.16 Dr D.J. Croome
University of Bath

In concert hall acoustics there are two clients – performers and listeners – and it must be remembered that conductors use the hall for rehearsals and some, such as Herbert von Karajan, use it for recordings. Over recent years it has become evident that the success of a concert hall does not depend on one parameter, such as reverberation time (although this should vary with frequency), but on a pattern of spatial and temporal factors, which include the shape of the sound build-up and decay curves, the shape of the concert hall and the distribution of sound throughout the space. In designing the acoustics the consultant should work with the musicians and, if possible, a sample of typical listeners.

A question first of all to Mr Day and Mr Middleton: can they throw any light at all on any work that has been done in large spaces such as pneumatic structures? I wonder if Mr Day could perhaps give a brief up-to-date account of the work in landscaped offices, because I think the Sabine formula, for example, does not work in these spaces. And a third question to Dr Barron: in his experience of theatres, can he say anything at all about noise insulation?

4.17 B. Day

We did a little bit of work a couple of years ago: a student was looking at the possibility of a temporary auditorium for the Eisteddfod, and we measured the acoustic reflection coefficient of air-backed canvas. One's immediate reaction is that canvas with just air behind it would be rather absorptive: in fact, the reflection coefficient was about 0.5. In other words, about half the energy incident on the canvas was reflected back. We subsequently did some measurements in the tent that was used for the 'Spirit of British Enterprise' exhibition held on the South Bank two or three years ago. Although, granted, it was quite a large tent, it had a reverberation time of about 1.5 sec at mid- to high frequencies. I think you would certainly need assisted resonance to provide a reasonable concert-hall acoustic for the low frequencies in a tent concert hall, but I was rather surprised at just how much reverberation there was at the high frequencies. One should perhaps not have been, because a heavy, densely woven canvas is quite a good membrane. The mass law is going to determine the degree of reflectance, as it is not a very porous structure.

I had quite a lot to do with landscaped offices at one stage, when we thought that we had discovered a way in which you could produce very rapid attenuation rates. It seemed to us that the build-up of noise in landscaped offices had a lot to do with how rapidly sound was attenuated with distance. The sort of things that Mr Middleton was saying about the uniform reverberant-sound fields not actually operating was absolutely true: the attenuation rate was about 3 dB for distance doubling. However, in a few offices we found very rapid attenuation. It dropped off even faster than in the open air at distances beyond four or five metres. We looked into the theoretical reasons for this and we came up with the sort of model I was talking about in my lecture. The theoretical model, which explained why, was developed just before we discovered that, in fact, having a rapid attenuation rate in the far field was just what you did not want. If you had that, it suppressed all the hundreds of conversations going on at a distance, which provided a nice gentle background noise to mask local sound. If you have this high attenuation rate, then it is very quiet and you can overhear your nearest neighbour. The kind of decay curve we were predicting gave perhaps 3 dB for distance doubling out to three or four metres and then maybe 9 dB for distance doubling beyond that; in fact, you want quite the reverse. You want very steep attenuation for local conversation and not much attenuation over large distances so that all the hundreds of conversations in the rest of the office would provide the 'noise perfume'. At that point we decided that as it was not a particularly useful thing to have, we would not investigate any further how it arose. You could always put the noise back in with electronic 'noise perfume', and that was really the last I had to do with landscaped offices.

Personally, I am a bit put off 'noise perfume' by the story that somebody tells about an office in which they installed it to solve the privacy problem. The system was installed without a switch but within 24 hours somebody had found a pair of wirecutters and cut the wires. The wiring was then put in conduit; within 48 hours someone had found a hacksaw and cut through the conduit. This story was told at a meeting to discuss the shape of the spectrum that 'noise perfume' ought to have; everybody laughed, but then went back to discussing the shape of the spectrum! I think that is a very irresponsible attitude. If an occupant is prepared to go to the trouble of cutting through a conduit to stop the noise, it is pretty obvious that they do not actually want it. 'Noise perfume' is not a way to solve the problem. In fact, I think landscaped offices probably should not be inflicted for that reason, unless you can provide adequate spacing between people, which is probably much more extravagant than we can afford to be. I do not think there is a real solution.

4.18 Dr M. Barron

I presume you do not want me to say anything about insulation as such, because we have heard people today who know much more about it than I do, but one recent discovery is that we are probably more sensitive to background noise in theatre environments than might have been thought. The normal recommended sound level in theatres is NC 20, which means that often we end up with NC 25 and this does, in fact, appear to interfere with hearing the spoken voice at the furthest seats in theatres, and a significant improvement in listening conditions can be achieved by getting that figure down to NC 20.

4.19 P.H. Andrew
Kennedy & Donkin

I just wanted to go back to a point that Dr Pinfold raised earlier. I think that the expression that he was after was 'antiphonal' effect rather than 'contrapuntal', when referring to old music in large buildings. In the days when this music was composed it was played mostly by bands or orchestras which were situated in different parts of the building, not all in front of the audience, and I think that his point as to where the music came from is relevant to that.

4.20 Dr M. Barron

I think what I was describing was preference. You seem to be describing a loudness control that quite a lot of amplifiers have or had, that gave a boost to the bass, which enables you to compensate for the frequency response of the ear when you are listening at a lower level. That seems perfectly normal because it is just a reflection of the equal loudness curves. This presence control was in addition to the normal treble and bass controls and, as I said, provided just a 5 dB boost or cut around 1 kHz. In spite of the smallness of the change, it was, nevertheless, clearly noticeable.

4.21 B. Day

I would like to add to the discussion with regard to the frequency response characteristic. It seems to me to represent rather a good illustration of my earlier point about the conflict between a full physical description of the properties of an auditorium and the way we perceive. If you take a loudspeaker radiating a pure tone and pick up the signal that is then produced at a microphone in the same room, and then gradually glide the tone up the frequency scale, the output of the microphone passes through a series of maxima and minima, perhaps 10 or 20 dB in height, spaced at perhaps 10 or 20 Hz. This kind of frequency-response test is often carried out on amplifiers, and one is unhappy with level variations greater than 3 dB; in an auditorium we just do not seem to be aware of them. They are there physically (and can be measured with an amplifier, loudspeaker and microphone) but the perceptual mechanism is able to smooth them out. It may be that we use the variations to tell us something about the subjective properties of the hall, but we are a long way from being able to take such measurements and relate them to perception.

CONFERENCE SUMMARY

Dr D.J. Croome

Dr Gilford reminded us that solutions to noise problems require an interdisciplinary approach. The energy age presses the building design team to consider economic solutions which will meet the needs of various users throughout the life of the building. A finer tolerance on heat and light, as well as the sound criteria, is being demanded. Noise emerges as both an art and a science in Chapters 1 and 2, by Dr Bryan and Professor Large, respectively. The social nature of the problem is reflected in the community indices such as the Noise and Number Index for aircraft, the L_{10} 18-hours one for traffic, the Corrected Noise Level for industrial sources and the integration of these into the unified index, Equivalent Continuous Sound Level, L_{eq}. Nevertheless, the individual response to noise is important and this introduces a tremendous variety of response to all the different types of noise emitted in different background situations. The susceptibility of people to their environment shows very large variations and the conclusions in those first two chapters are that physiological sensitivity, psychological sensitivity, arousal level and distraction are cardinal dimensions of environmental response.

Mr Taylor, Mr Edwards and I showed planning and design approaches to sound control in buildings. Common sense and good workmanship, as well as technical knowledge, are all important if the problems encountered in the case studies described are to be avoided. Communication between manufacturers, designers, contractors and builders is paramount. We heard from Mr Edwards that some clients were not happy even with NR 25, and Dr Mackenzie feels designers should use the PNC criteria if low-frequency rumble is to be avoided, rather than NC or NR ratings. Each site needs appraising and, wherever possible, the client should listen to such things as fan-coil units, or other systems equipment similarly installed in other buildings, before a brief and specification are fixed. It appears that building services consultants have no alternative at the moment but to over-design to ensure clients' satisfaction. Tradition can stifle progress, and evidence was put forward in the chapters by Mr Redmore and myself which questions the validity of certain design calculation procedures. Much reliance is put on data which are being measured for individual elements of a system, whereas in practice the total system does not react as the sum of the parts. Interaction between the elements is important and sound generation at low velocities can offset natural attenuation, especially in systems having a high proportion of fittings.

Some latest developments were reviewed by Mr Spring. These covered rigid absorbers, laminated glass windows, weather-stripping, high-damping polymers and microprocessor application such as active attenuation and the use of microcomputers – which present an opportunity to solve acoustic problems, but I think there is a note of caution to be added here. There are not any magic solutions, and ideas such as anti-noise may be helpful in some special circumstances. Dr Mackenzie states that if design aids are to be effective, equipment manufacturers need to provide more information about sound, power and vibration levels, and designers should make tighter specifications concerning workmanship detailing. Improved communication and education at all levels are clearly required. The design and research approach to a sound-insulation case study for a housing development scheme was described by Mr Box on behalf of Mr Simpson. Besides site planning for noise control, performance tests of a specially designed window and airbrick were undertaken, and it is interesting to note that it is intended that the tenants will be interviewed eventually to establish whether the sound-control measures taken will be effective in human terms.

Conceptual mathematical and physical models simulating behaviour of sound in spaces have been evaluated for us by Mr Day. Their application in the design process can be invaluable, especially in the case of large, open spaces, such as concert halls and multi-purpose theatres. The Health and Safety at Work Act of 1974 has emphasised the importance of ensuring good working conditions throughout industry and Mr Middleton has explained how layouts, sound-absorption barriers, partitions and enclosures can be used which can alleviate the problems of sound exposure for the employees. Ways of minimising the noise emission from industrial premises were also given.

Designing spaces for music is perhaps one of the most interesting aspects of acoustics and one in which there is a great interplay between architecture, engineering and science. I think that the art of this probably lies in achieving this interplay. Dr Barron has explained how progress has developed, and concentrated on the range of subjective characteristics required for good acoustics. Spatial response is an important feature in the visual aspects of environmental design as well as the aural ones. The problem of achieving a low sound level is, of course, critical in a concert hall. Assisted resonance provides an electronic means of achieving flexibility in multi-purpose halls.

The discussion periods have been extremely valuable and have affirmed that noise is a critical factor in the design of buildings and services. Problems arise mainly from a lack of common sense, or technical knowledge, and poor workmanship and communication. Perhaps our conclusions might be that better education in acoustics is needed not only for architects and engineers, but also at the operative level, because it is not only the science and art which are important, but also the craft of building acoustics too. Manufacturers, contractors and designers should co-operate more. There needs to be more feed-back data obtained from buildings in use, about equipment as well as human requirements. It is hoped that this conference may be judged to be a success because it has emphasised these needs and has indicated how some problems may be overcome. More research is needed, but a better application of existing knowledge is equally important.

I should like to finish by thanking the conference organisers, CICC, the Technical Organising Committee, the authors, the speakers and last, but not least of course, yourselves for being such an interesting audience.

Index